Marketing to the Base of the Pyramid

Ramendra Singh · Tahir Ahmad Wani
Editors

Marketing to the Base of the Pyramid

Towards a Sustainable Business Future

Springer

Editors
Ramendra Singh
Indian Institute of Management (IIM) Calcutta
Kolkata, West Bengal, India

Tahir Ahmad Wani
Department of Humanities Social Sciences and Management
National Institute of Technology (NIT)
Srinagar, Jammu and Kashmir, India

ISBN 978-981-96-7104-5 ISBN 978-981-96-7105-2 (eBook)
https://doi.org/10.1007/978-981-96-7105-2

This Springer imprint is published by the registered company Springer Nature Singapore Pte Ltd.
The registered company address is: 152 Beach Road, #21-01/04 Gateway East, Singapore 189721, Singapore

We wish to dedicate this edited book to all management professionals, policy makers and management educators who sincerely believe in making the world a better place to live for one and all, beyond paying the usual lip service.

Dr. Ramendra Singh and Dr. Tahir Ahmad Wani

Preface

This edited volume investigates the convergence of sustainability and marketing at the Bottom/Base of the Pyramid (BoP), contesting the notion that sustainability is a notional concept that is relevant for the extravagant environments. Linking sustainability to the foundational aspects of marketing elucidates the potential for peaceful coexistence between the two contexts. This book, featuring contributions from respected scholars in BoP and subsistence markets, presents innovative viewpoints on how sustainable practices can meet the distinct requirements of low-income customers while promoting economic, social, and environmental welfare. Markets at the Bottom of the Pyramid are characterized by resource scarcity, fragile infrastructure, and distinctive socio-cultural contexts. Sustainability and Marketing to the Base of the Pyramid delves into how businesses, policymakers, and communities can overcome these challenges to create inclusive and impactful solutions. It examines innovative approaches to product and service design that cater to the aspirations of BoP consumers, emphasizing cultural alignment and community collaboration.

The book contains a wealth of information demonstrating the importance of economic sustainability in BoP marketplaces, followed by social and environmental factors, in addressing serious disparities and resource constraints. The role of collaboration between corporations and communities in combating climate change is often highlighted in all chapters. The key insights range from discussion on the transformative potential of marketplace literacy programs to empower BoP consumers, or insights into inclusive marketing practices that foster well-being in low-income communities by avoiding common pitfalls and building socially and environmentally balanced marketplace exchanges. The ideas also include exploring frugal, grassroots, and reverse innovations that enable BoP consumers to overcome constraints and achieve sustainable growth. The book also highlights the complexities of sustainable consumer behavior in emerging markets, features innovative role of intermediaries, such as JUNTOS (in Argentina), in bridging informal and formal markets, and addressing the market habitus gap. Other chapters include discussion on the strategies employed by luxury brands to involve marginalized BoP communities in value creation, thereby enhancing their well-being while redefining the role of high-end brands in social development. Yet another chapter highlights the importance of a bottom-up approach to understanding subsistence marketplaces, offering unique insights into sustainable development through product innovation, education, and social enterprises. In a time that is brimming with too many global challenges, this book serves as a valuable resource for understanding how businesses can balance profitability with social impact in underserved markets. It offers practical recommendations for academics, policymakers, and business leaders to create innovative, sustainable, and inclusive solutions for BoP markets worldwide.

As Editors, we suggest that it may be an ideal companion for students, policymakers, and researchers in marketing, business administration, rural management, economics, and development studies, as it shows a roadmap for creating a more equitable and sustainable future through the integration of marketing and sustainability at the Bottom of the Pyramid.

The book has 9 chapters. All chapters are unique in their own way. We briefly discuss the key insights offered by each chapter below.

Chapter 1 titled, "A Market Void: Are We Missing a Market Actor? Insights from JUNTOS's Action-Research Case," investigates the sustainability challenges in critical social dimensions. The action-research case looks at how to enable bottom-of-the-pyramid (BOP) or informal sellers to market their products beyond local fairs and community consumers, which is aligned to United Nations' 17 Sustainable Development Goals. A failed pilot project aimed at integrating informal sellers into multinational corporation (MNC) supply chains and a successful subsequent action-research case, JUNTOS, provided insights and solutions into the challenges faced by informal sellers in accessing formal markets. The chapter delves into key learnings that emerged from failed pilot project, and how reframing the problem from a regulative-type to a cognitive-cultural-type definition revealed the existence of a missing market actor. This insight led to the development of JUNTOS as a permanent intermediary able to bridge the formal/informal market habitus gap. Leading to the learnings and post-successful implementation, JUNTOS has grown and thus is a living proof of the efficacy of the model. The chapter argues that the creation of such permanent intermediaries can address the market-actor void, offering a sustainable solution for informal sellers to access formal markets.

Chapter 2 titled, "Inclusive Marketing: Fostering Sustainable Wellbeing in Subsistence Communities," looks at how the key challenges before marketers are to cater to the underserved needs of low-income consumers around the globe by ensuring that people from low-income contexts have access to quality products and services. In this chapter, the authors look from an inclusive marketing mindset to help address the unserved needs of creating new marketplace exchanges where none existed or by supplanting a less optimal marketplace exchange with one that balances the process of creating social, environmental, and economic value. The authors highlight eight mistakes commonly made by marketers who seek to improve the well-being among low-income communities in a sustainable manner, and also make recommendations to the marketing practitioners to go for inclusive and sustainable practices to address the underserved needs in the low resource contexts.

Chapter 3 titled, "Marketplace Literacy and Sustainable Consumption at the Base of the Pyramid (BoP)," explores the transformative role of marketplace literacy programs in fostering sustainable consumption and empowering consumers at the Base of the Pyramid where the local communities face distinct socio-economic and cognitive challenges, including limited literacy, income constraints, and social pressures. The author argues for marketplace literacy programs that can address these challenges and enable BoP consumers to make informed, value-oriented choices. The author then integrates theoretical and empirical insights to highlight how marketplace literacy aligns with Sustainable Development Goals (SDGs), particularly SDG 1 (No Poverty), SDG 4 (Quality Education), and SDG 12 (Responsible Consumption and Production). The author also shows how marketplace literacy fosters resilience by building metacognitive and social intelligence skills, while the role of ICT-enabled interventions in bridging information asymmetries, empowering women entrepreneurs, and promoting sustainable digital consumption, is also highlighted.

Chapter 4 titled, "Bridging Economic, Social, and Environmental Sustainability in BoP Markets: An Integrated Approach," adapts Lim's Sustainability Pyramid approach to BoP contexts and proposes that sustainability initiatives in BoP markets should prioritize economic sustainability, followed by social and environmental dimensions. The authors suggest that this hierarchical approach aligns with BoP realities, where economic concerns often take precedence. The chapter further explores how academics, policymakers, and firms can collaboratively influence consumer decision-making towards sustainable consumption in BoP markets using this adapted framework, which has the potential to drive meaningful, long-term sustainable development in BoP communities.

Chapter 5 titled, "Bridging Vulnerability and Sustainability: Collective Climate Change Adaptation Efforts by Indian Communities and Businesses," argues that for emerging markets the threat of climate change is greater than for other nations as the dependence on natural resources for survival of BoP communities is quite high. The authors examine the inter- connectedness of climate change and poverty targeting bottom-of-the-pyramid communities. The BoP communities often deal with both natural dangers with limited availability of resources that make the challenge severely difficult. The chapter further evaluates various coping strategies, including technological solutions, agricultural innovations, and migration, changing cropping patterns, borrowing and financial support, livestock management, utilizing natural resources as a coping mechanism. The chapter also highlights the role of business organizations in managing climate change and consequently, highlighting the social and financial challenges businesses face when adopting climate-friendly procedures due to the possibility of unwanted outcomes caused by extensive interventions.

Chapter 6 titled, "Sustainability in Subsistence Marketplaces and Beyond: Addressing Challenges Through the Bottom-Up Approach," presents a comprehensive cumulative picture of the body of work conducted by the author and his cohort for previous two and a half decades with low-literate, low-income consumers, entrepreneurs, and marketplaces—i.e., subsistence marketplaces. A comprehensive overview of key insights is presented that have emerged at the intersection of sustainability and subsistence. A bottom-up perspective is highlighted, beginning at the micro-level behavioral understanding of subsistence marketplaces as they intersect with the environment, consumption and conservation, in contrast with a top-down perspective.

Chapter 7 titled, "Empowering the Bottom of the Pyramid: Harnessing Innovation for Sustainable Growth," looks at the various aspects of BOP-focused innovations, such as grassroots innovation, reverse innovation, and Jugged innovation, that can empower the BOP with social, economic, and environmental sustainability. The authors draw on theories such as the resource-based view, community-based theory, theory of disruptive innovations, and effectuation theory to support their views. The key insight being that how each form of innovation can enhance the lives of individuals in low-income settings. The authors support their ideas by providing real-world examples, and draw attention to the role of an enabling ecosystem that includes supportive government policies, accessible infrastructure, access to finance, robust social networks, and technology transfer. These enabling factors as authors argue can help expedite the innovation processes and foster sustainable and inclusive growth.

Chapter 8 titled, "Sustainable Luxury Brands Strategies for Base of the Pyramid Wellbeing," argues that while luxury brands are producing high-quality products with an aspirational status for non-affluent customers, it is leading to higher distrust of consumers from marginalized sections of the society. The authors explore various innovative strategies implemented by the luxury/premium brands for the BOP communities' well-being. These strategies transcend the brand-consumer transactional relationships. The authors argue that the BOP wellbeing is catered by inclusion of these communities as suppliers and employees to create value for the end consumers, which directly impacts the daily lives of BOP communities. Four different luxury brands (FabIndia, Jaipur Rugs, Farm2kitchen, and Sabyasachi) are analysed to understand their strategies and tactics for the BOP wellbeing and inclusion. The integration of theories with brands' data helped the authors to suggest five distinct strategies for BOP wellbeing.

Chapter 9 titled, "Sustainable Consumer Behaviour in Emerging Markets," looks at the sustainable consumer behaviour which refers to the extent to which consumption decisions are undertaken with the intention of reducing or improving the effects on society and the environment. The author challenges readers into questioning whether the global markets are assumed to be socially and environmentally conscious of how humans impact the environment. The author argues that the responses and expectations of consumers may vary depending on contextual variables of their region, and go on to discuss the dimensions and complexities of sustainable consumer behaviour including segmentation characteristics, the predictors, and willingness to pay with a special focus on consumers in emerging markets at the BoP. The chapter also proposes a comprehensive description of sustainable consumption behaviours and highlights the relevance to environment and social-marketers who wish to target consumers and design intervention strategies for this target group.

As we present the edited volume to the readers, we wish that we are able to slightly move the mountain of knowledge and hill of management practice towards making this world slightly more sustainable especially for the marginalized sections of our society.

Kolkata, India Ramendra Singh
Srinagar, India Tahir Ahmad Wani

Contents

Editors and Contributors

About the Editors

Ramendra Singh is Professor of Marketing at IIM Calcutta. He completed PhD from IIM Ahmedabad, MBA from XLRI Jamshedpur, and B.Tech. from IIT-BHU. His research has been published in leading international journals including, *Journal of Business Research, International Marketing Review, Marketing Theory, Industrial Marketing Management, Journal of Personal Selling and Sales Management*, among others. He has previously worked in sales and marketing responsibilities across different industries for several years. He has authored a textbook on 'Sales and Distribution Management' which has been adopted in MBA programs at IIMs and other business schools. He has completed several consultancy projects, and conducted several training programs for senior and middle management cadre at several organizations.

Tahir Ahmad Wani is Assistant Professor in the Department of Humanities, Social Sciences and Management, National Institute of Technology (NIT) Srinagar, India, wherein he teaches various management-related subjects to management and engineering students. Before joining NIT Srinagar, he was working at School of Business Studies, University of Kashmir, India. He has a rich research background and has written extensively on various management issues. He has also conducted workshops on quantitative analysis and participated in numerous workshops and conferences in reputed institutions across India.

Contributors

Anaka Aiyar University of Vermont, Burlington, VT, US

Amrita Bansal IMI, New Delhi, India

Neha Bhardwaj Faculty of Management Studies, University of Delhi, Delhi, India

Shaphali Gupta Indian Institute of Management, Kozhikode, Kozhikode, India

Saravana Jaikumar IIM Calcutta, Kolkata, India

Varsha Jain MICA, Ahmedabad, India

Mahima Mathur Department of Retail Management, Prin. L. N. Welingkar Institute of Management Development & Research, Mumbai, India

Falaq Mubashir HSS&M, National Institute of Technology Srinagar, Srinagar, India

Jaqueline Pels Universidad Torcuato Di Tella, Buenos Aires, Argentina

Ramendra Singh IIM, Calcutta, India

Rushabh Trivedi MICA, Ahmedabad, India

Srinivas Venugopal University of Vermont, Burlington, VT, US

Prama Vishnoi Keshav Mahavidyalaya, University of Delhi, Delhi, India

Madhu Viswanathan Department of Marketing, College of Business Administration, Loyola Marymount University, Los Angeles, CA, USA;
Department of Business Administration, Gies College of Business, University of Illinois, Urbana-Champaign, Champaign, IL, USA;
Nottingham University Business School, Nottingham, UK

Anupama Vohra Faculty of Management Studies, University of Delhi, Delhi, India

Tahir Ahmad Wani HSS&M, National Institute of Technology Srinagar, Srinagar, India

Chapter 1
A Market Void: Are We Missing a Market Actor? Insights from JUNTOS's Action-Research Case

Jaqueline Pels

Abstract Sustainability challenges include critical social dimensions. One unresolved issue is how to enable bottom-of-the-pyramid (BOP) or informal sellers to market their products beyond local fairs and community consumers. This concern is reflected in several of the United Nations' 17 Sustainable Development Goals. A failed pilot project aimed at integrating informal sellers into multinational corporation (MNC) supply chains and a successful subsequent action-research case, JUNTOS, provided insights and solutions into the challenges faced by informal sellers in accessing formal markets. Drawing on the pilot project analysis and an extensive literature review, the chapter underscores that, traditional solutions focus on helping informal sellers to transition from informal to formal markets by providing training, and micro-lending. This approach has been labeled as a regulative-type approach; i.e., learning new rules. However, this approach often fails due to overlooking a deeper issue: the market habitus gap. This gap, rooted in distinct cultural and cognitive dispositions between informal and formal market actors, hinders informal sellers' access to formal markets. The literature review also revealed that the pilot case was no exception and that most current regulative-type solutions, involving temporary intermediaries that provide mentoring and financing, have not proven successful. Drawing on Pierre Bourdieu's theory of habitus, the paper emphasizes that informal and formal markets operate under different sets of practices, necessitating a new approach. Adopting a cognitive-cultural perspective highlights that the market habitus gap not only poses a larger challenge than previously recognized but also impedes informal sellers from effectively accessing formal markets. Reframing the problem from a regulative-type to a cognitive-cultural-type definition revealed the existence of a "market-actor void," meaning that a necessary market actor is missing. This insight led to the development of JUNTOS as a permanent intermediary able to bridge the formal/informal market habitus gap. Unlike traditional mentors or financial institutions, JUNTOS actively facilitates trade between informal sellers and formal buyers by providing both bridging social capital and cultural capital in the form of marketing and managerial capabilities. JUNTOS operates as a non-digital marketplace to ease

J. Pels (✉)
Universidad Torcuato Di Tella, Buenos Aires, Argentina
e-mail: jpels@utdt.edu

R. Singh and T. A. Wani (eds.), *Marketing to the Base of the Pyramid*,
https://doi.org/10.1007/978-981-96-7105-2_1

the transactions. In short, JUNTOS allows to overcome challenges such as lack of social networking, working capital, and market literacy. Over six years, JUNTOS has grown significantly, demonstrating the effectiveness of its model. The paper argues that the creation of such permanent intermediaries can address the market-actor void, offering a sustainable solution for informal sellers to access formal markets. This model contributes to inclusive economic development and has implications for public policy, particularly in rethinking support mechanisms for informal sellers.

Keywords Informal sellers · Formal and informal markets · Market voids · Market habitus gap · Cognitive-cultural institutional literature

1.1 Introduction

Sustainability urgencies go beyond environmental issues (Brundtland, 1987); they involve all 17 of the United Nations' Sustainable Development Goals or SDGs (United Nations, n.d.). In this chapter, I focus on the social dimensions of sustainability; specifically, on informal sellers' access to markets.

While reaching the bottom of the pyramid (BOP) with quality products and services is certainly important (Prahalad, 2004), equally important is enabling sellers from the BOP to sell their products beyond local fairs and community consumers (Pels et al., 2022). Little and Dolan (2000) argue that selling in informal markets hinders efforts to lift people out of poverty, while McMullen (2011, p. 188) states that operating in informal markets lead to "high opportunity costs by preventing gains from trade with strangers". To overcome these difficulties, informal sellers need to trade in formal markets. However, as Hamlet said in Act 3.1.64, "Ay, there's the rub" (Shakespeare, 1603/n.d.).

The problem of enabling informal sellers to gain access to formal markets is pressing because in emerging markets between 30 and 70 percent of all businesses are unregistered (International Labour Organization, ILO, 2023). To discuss the complexity of enabling informal sellers to sell in formal markets, in this chapter, I draw on an action-research project (Avison et al., 1999) titled *JUNTOS: Compras con 100% Impacto Social* (Purchases with 100% Social Impact). The value of using illustrative cases like JUNTOS is that they help make conceptual work and abstract constructs more tangible and accessible (Siggelkow, 2007). The JUNTOS project was carried out by the Inclusive Business Think Tank (ENI,[1] for its Spanish acronym) at the Business School of Universidad Torcuato Di Tella in Buenos Aires. As the director of the ENI team, I have been actively involved in the JUNTOS initiative, and I also incorporate insights from papers co-authored with international colleagues.

The structure of this chapter follows the three phases of JUNTOS's action-research process, an approach that allows a blending of case detail and theoretical constructs. To interpret the data and identify novel concepts elicited from JUNTOS, a theorizing

[1] https://www.utdt.edu/ver_contenido.php?id_contenido=11120&id_item_menu=21522.

process based on abduction is adopted. Abduction allows us to associate "what … [we] see in the empirical world with theoretical ideas, which are also out there and can be further developed" (Gehman et al., 2018, p. 297).

The chapter offers several contributions. First, it demonstrates, contrary to understandings in mainstream literature, that very few informal sellers manage to *transition* into selling to formal markets. Second, it identifies a market habitus gap that presents a greater challenge for these sellers than is currently recognized. This gap impedes informal sellers from accessing formal markets, thereby resulting in a market void. Finally, it advocates that overcoming the market void requires a new market actor—a new permanent intermediary that can act as a bridge between formal and informal markets by providing both contacts (i.e., bridging social capital) and marketing and managerial capabilities (i.e., cultural capital).

1.2 Phase 1: A Failed Pilot Project

In July 2014, an association of multinational corporations (MNCs) invited the ENI to study and evaluate an inclusive supply-chain pilot program. If the results of the 18-month pilot were satisfactory, the program would be replicated in the 11 Latin American countries where the set of MNCs was operating. Four of the MNCs participated in the pilot program, which involved buying non-strategic products from local BOP sellers.[2] The program had two complementary goals. The first was to equip local informal sellers with the skills needed to sell to MNCs. The second was to help the MNCs develop knowledge of how to acquire goods from local informal sellers. At the ENI, we saw the program as an opportunity to learn more about inclusive supply chains and appreciated that its aims aligned with Little and Dolan's (2000) and McMullen's (2011) comments that selling to formal markets is important. Moreover, if successful and rolled out to the rest of Latin America, this initiative could potentially reduce the informal market sector, which accounts for an average of 30 percent of the gross domestic product in developing economies (Loayza, 2018). The project was aligned with mainstream thinking, that positions mentoring, training, and providing financial support as the key instruments in helping informal sellers transition between markets (Sutter et al., 2017; Yunus, 1999).

The four pilot MCNs formed an alliance with an NGO dedicated to mentoring and training subsistence entrepreneurs in technical and managerial skills. One of the MCNs also acted as a financial unit for the project, thus simplifying the process of accessing working capital. Additionally, every six months, we interviewed all actors (the BOP sellers, the MNCs' clients, the mentoring NGO that provided technical assistance, and the MNC that acted as the financial unit). We also registered the whole project so that we could analyze the evolution and performance of the pilot.

[2] The terms "informal sellers" and "BOP sellers" are used interchangeably.

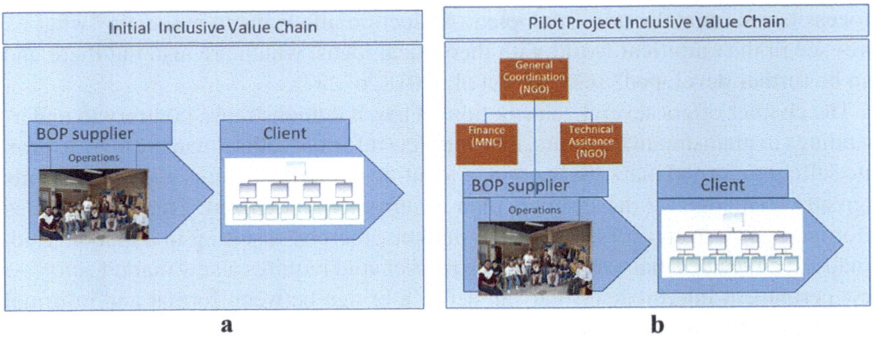

Fig. 1.1 **a.** Inclusive value chain prior to the pilot project **b.** Inclusive value chain during the pilot project

In our final report on the pilot project,[3] we included a set of illustrative images that show the support provided to the BOP suppliers. Figure 1.1 illustrates the structure of the inclusive value chain prior to the pilot program, while Fig. 1.1 depicts the more complex inclusive value chain developed during the pilot project.

By December 2016, although the pilot had achieved some of its operational goals, none of the BOP sellers expanded beyond working with a single MNC or did they become independent from the mentoring NGO. As a consequence, the project was not expanded to the remaining Latin American countries.

1.3 Phase 2: Searching for Answers

We were frustrated and perplexed when the pilot program failed to meet its objectives. It had followed the mainstream tips on how to help informal sellers' transition to formal markets and provided numerous resources. So, what had gone wrong?

1.3.1 Was It an Exception?

Had the mainstream literature overlooked something? Effort to answer these questions resulted in an article titled "Informal Sellers and Formal Markets: A Habitus Gap" (Pels et al., 2022). It reviewed the literature that sought to overcome the challenges faced by informal sellers to operate in formal markets. The article concluded that the challenges remained unresolved, a finding reinforced by the failure of the pilot program. Pels et al., (2022, p. 1270) highlight that:

[3] 2014–2015 | Programa Generar https://www.utdt.edu/ver_contenido.php?id_contenido=11125&id_item_menu=21531.

> Policymakers focused on: (1) aiding the poor capitalize on or develop entrepreneurial capabilities through 'institutional scaffolding'; i.e., training and mentoring (Sutter et al., 2017) to help them sell their goods/services, and (2) providing financial assistance in the form of innovative micro-lending initiatives (e.g., Yunus, 1999).

Pels et al. (2022) further conclude that:

> Studies on the impact of training/mentoring show that, although interventions have tended to work with carefully selected populations, the overall results have been disappointing (e.g., Kistruck et al., 2013; Sutter et al., 2017). Similarly, studies on micro-credit performance show that "…most new firms, funded by micro-credits agencies, start as self-employment efforts, and remain so" (Bradley et al., 2011, p. 644).

The ENI team concluded that the pilot program's disappointing outcome was not an exception but rather the norm, prompting us to explore the literature for a different approach to addressing "the rub". The review yielded several insights, the first of which compelled us to redefine the problem.

1.3.2 Insights from the Literature Review

A first insight came from the institutional literature. Scott (2014) classified this literature into three pillars—regulative, normative, and cultural-cognitive. The regulative pillar relates to formal and informal rule-setting, monitoring, and sanctioning activities. The normative pillar, a prescriptive dimension in social behavior, defines what is appropriate and reasonable conduct in a particular social setting, while the cultural-cognitive pillar concerns the emergence and use of the common schemas and shared mental models that guide behavior and produce stabilizing effects on social structures.

Pels et al. (2022) understood that the gap between informal/formal markets went beyond the regulative distinction and advocated shifting the problem definition from an institutionalist regulative type of problem to a cultural-cognitive type of problem. According to Scott (2014), cultural-cognitive theorists amend and expand the representation of institutions proposed by regulative-inclined academics. Theorists holding cultural-cognitive views argue that markets involve more than rules and enforcement mechanisms because sellers and buyers participate in markets endowed with different capacities for action. This shift implied a need to modify how formal/informal actors perceive markets.

The regulative approach, following North (1990), focuses on markets as institutions governed by formal and informal rules and poses the *problem as one of transitioning* between markets.[4] The literature review also allowed to identify the

[4] *Formal institutions* include laws and regulations (i.e., property rights, contract law) as well as the supporting apparatuses that monitor and enforce actions within society's legal boundaries. *Informal institutions* include the norms, values, and beliefs that shape notions of social acceptability and provide some localized structures to support economic activity.

institutional regulative type of problem definition as the one adopted in the pilot project. Alternatively, the cultural-cognitive approach, invites to.

> conceive formal and informal markets as Bourdieusian fields governed by different dispositions which govern not just action but perception, understanding, motivation and so on. In short, formal and informal markets are characterized by different habitus and for an informal seller to transition to a formal market involves the mastery of a set of practices based on different dispositions. We call this bridging a market habitus gap. (Pels et al., 2022, p. 1271)

This conception implied the need to *shift the problem definition from transitioning to accessing,* which renders the problem a marketing one.

Introducing the Bourdieusian notion of habitus involves (briefly) unpacking the three types of capital—economic/financial, cultural, and social (Bourdieu, 1986, 1999). Economic capital refers to the assets that individuals possess. Encompassing wealth, income, and property, it shapes opportunities and inequalities within societies, while financial capital is "the direct or indirect mastery (through access to the banks) of financial resources" (Bourdieu, 2005, p. 194). Social capital refers to the resources and benefits individuals gain through their social networks, relationships, and group memberships. It includes the connections and mutual obligations that provide access to opportunities, information, and support. Cultural capital denotes the non-financial assets that individuals acquire, such as education, skills, knowledge, and cultural tastes, all of which influences their social mobility and status. Bourdieu identifies three forms of cultural capital: embodied (e.g., traits and knowledge), objectified (e.g., cultural goods such as books and art), and institutionalized (e.g., academic credentials).

Bourdieu (1989) coined the concept of habitus to emphasize that the three forms of capital are intimately bounded and perpetuate social stratification and hierarchies. More importantly, Bourdieu (1989, p. 19) referred to habitus as "a system of schemes of production of practices". Thus, habitus can be understood as the "space of possibilities" open to a person, who is conditioned by the three types of capital they possess. Central to my argument in this chapter is the durability of the disposition of habitus. Habitus "contains a tendency to persist in its being, is a force inscribed in the objectivity of things so that everything is not equally possible or impossible" (Bourdieu, 1986, p. 42). For example, informal markets often have no fixed price and deadlines are flexible. While these traits may seem easy ones to learn, they are deeply rooted and reflect the broader cultural-cognitive setting informal sellers are embedded in. More specifically, these forms of understanding exchange relate to the flexibility needed to overcome the beyond-business personal and situational circumstances informal sellers usually encounter. Consequently, modifying these behaviors implies changing life-surviving patterns.

A second insight emerged from exploring the literature on subsistence entrepreneurship. It signaled that most informal sellers are not entrepreneurs in the traditional sense; rather, they are essentially production units—individuals who possess a particular craft (e.g., sewing) but have no marketplace literacy (Viswanathan et al., 2009). These people generally also lack the basic accounting and financial skills needed to factor in the costs of pursuing economic opportunities (Banerjee, 2013;

Viswanathan et al., 2009). London et al. (2010) label these skill deficits as transactional constraints on capturing value. Pels et al. (2022) again drew on Bourdieu (1999) and his notion of embodied capitals, that is, the capitals acquired from childhood through a continuous process of socialization over a substantial period of time and not instantaneously transmissible. Embodied capitals are relevant because they present as a "marker of life chances and opportunities" (Bourdieu, 1999, p. 22).

In keeping with this view, several authors in the literature stress that the role of human capital is essential to entrepreneurial activity at the BOP and that human capital deficit prevents BOPers from engaging in the entrepreneurial activities which generate real economic growth (Bruton et al., 2011; Viswanathan et al., 2009; Webb et al., 2010). These views of human capital led Pels et al., (2022, p. 1271) to define "the informal seller as: (i) a self-employed person or group of people (usually forced into this situation through a lack of employment options); (ii) who develops a commercial activity based on an available skill or resource but with deficient managerial skills and limited capacity to detect market opportunities, and: (iii) ends up trading within an informal community". This characterization, we concluded, reflected the profile of the BOP sellers in the pilot project.

A third insight arose out of the scrutiny of institutional entrepreneurial literature (DiMaggio, 1988). Both the institutional intermediaries,[5] as well as the social intermediaries[6] were associated to the regulative type of problem definition and, therefore, solutions that tried to help informal sellers to *transition* between markets through mentoring and financing solutions. Sutter et al. (2017) maintain that the transition from informal to formal markets may involve a process that is more complex than a simple understanding of the rationale behind cost/benefit analyses. In their words, "transitioning from informal to formal markets involves fundamentally changing the way the business operates […] (and it implies) adopting a whole new mindset and knowledge base" (p. 421).

Turning again to Bourdieu (1986), Pels et al. (2022) considered his notion of capital conversion, that is, the effort required to convert cultural capital (such as training) into economic capital (such as gains from selling within formal markets). Pels et al. (2022) realized that the time/effort ratio of this conversion is typically too high, which explains the high failure rates (Karlan & Valdivia, 2011). Moreover, as prior studies show, even when some informal sellers manage to transition to formal markets, most slip back to informal markets at different stages of the transition process (Sutter et al., 2017). Pels et al. (2022) concluded that these disappointing outcomes could be explained by the above-mentioned habitus tendency to persist in its being (Bourdieu, 1986) and that the temporary nature of institutional/social intermediaries is part of the problem because they underestimate the time and effort required to achieve the capital conversion from production skills to managerial knowledge to sales. Once again, these insights gained from the literature helped us

[5] Usually, non-government organizations, working across institutional boundaries to facilitate entrepreneurs' transition from one institutional arrangement to another, Dutt et al., 2016.

[6] Usually, a type of social enterprise engaged assisting BOP producer in transition from one regime to another by developing acceptable transactional capabilities, Kistruck et al., 2013.

explain the pilot project results and, particularly, why the work of the NGO did not yield the expected outcome.

A fourth insight came from the realization that the literature uses the term social capital without specifying whether it is "bonding" or "bridging". These terms had their genesis in Granovetter's (1977, 1985) distinction between the "strong" and "weak" ties of networks and tie-formation based on the cultural-cognitive closeness of members. Gittell and Vidal (1998), in turn, associated Granovetter's ties with social capital and assigned these terms the labels of bonding and bridging. Bonding social ties form the core network, which is often the primary source of interpersonal contact. Bridging social ties are extra-community linkages that usually have weaker ties. Although bonding social capital is central in microcredit literature, bridging social capital is crucial for accessing formal markets. As Woolcock (2001, p. 11) puts it, the poor need bonding capital "to get by" and bridging capital "to get ahead".

Finally, Pels et al (2022) concluded that if BOP suppliers operating in informal markets are to sell to formal markets, then a novel solution as to how they can best do this is required. They accordingly proposed *that a new permanent market actor* is needed to help the sellers access these markets. The idea that the market is incomplete and that a new actor is needed was indirectly addressed by Yunus (1999) and his microcredit institutions. Yunus never tried to convince banks to have a special queue for microcredit borrowers (as they have for elderly people or pregnant women); nor did he try to teach microcredit borrowers how to ask for a loan or credit. Instead, he realized that what was needed was a new actor who could talk to both the financial world (to capture capital) and to the microcredit borrowers. Arguing along similar lines, Pels et al. (2022) suggested that a new permanent intermediary is needed to bridge the market habitus gap. This novel intermediary would differ from extant forms of intermediaries (institutional and social) by assisting BOP sellers. Essentially, this intermediary would reduce capital conversion costs by giving informal sellers access to new (formal) clients (i.e., bridging social capital) and providing the marketing and managerial capabilities (i.e., cultural capital) informal sellers lack.

Figure 1.2 presents a summary of the concepts developed so far. It positions the literature referred to when addressing these topics, illustrates the market habitus gap, and shows the position of the new intermediary, namely, JUNTOS, described in detail in the next section of this chapter.

1.4 Phase 3: A New Market Actor: JUNTOS Purchases with 100% Social Impact

The insights from the literature review by Pels et al. (2022) prompted us, at ENI, to formulate a new problem–solution definition centered on the existence of a market void and the need for a permanent intermediary actor. This led to the creation of JUNTOS Purchases with 100% Social Impact (hereafter referred to as JUNTOS).

This development enabled our ENI action-research project team to generate a set of hypotheses that we aimed to validate in 2018.

JUNTOS is a new permanent intermediary.[7] Unlike institutional and social intermediaries, JUNTOS neither mentors nor helps informal sellers' transition between markets. This new actor respects informal sellers' market habitus gap and reduces capital conversion costs by applying JUNTOS's managerial knowledge and capabilities and bridging social capital. The JUNTOS project nonetheless presented those of us in the ENI team with several initial challenges, evident in the questions we asked ourselves.

The first was *where to find the informal sellers?* Public databases of cooperatives (the main legal form these organizations adopt) made it possible to start identifying and visiting them. When we did, we identified several different types. For example, some had been established because of job loss, others due to the difficulty of gaining inclusion in the formal labor market, maybe because the people involved in them had spent time in prison or been unable to finish their schooling. This variety meant we had to develop a set of criteria to evaluate the cooperatives' potential for selling to formal markets. The criteria included, amongst others, what equipment they had, if they managed stock, and if they were flexible in their production capabilities. Common factors across all, however, were their restricted ability to produce large volumes and the well-known lack of working capital.

Our next question was *who might the informal sellers sell to?* Through our understanding of the capabilities of and restrictions on informal sellers, our JUNTOS team decided that MNCs (such as those in the pilot project) are not the natural buyer of goods from cooperatives. First, MNCs demand large volumes of products per order, which means informal sellers have to abandon other customers so they can allocate all their resources (people and equipment) to filling a single order. Second, due to their large volumes, MNCs usually buy from large producers whose economies of scale and scope enable them to offer lower prices and can offer significant discounts, thus outpricing the informal sellers. Third, large orders require access to additional working capital. Fourth, MNCs have strict (usually international) paying procedures, with payment normally occurring 30/60/90 days after product delivery. This practice places an even heavier financial burden on informal sellers. Fifth, MNCs require their suppliers to register on a suppliers' digital platform. However, most informal sellers do not have the skills for this or the platforms do not feature categories (e.g., cooperatives) inhibiting the informal sellers to register as suppliers.

We decided that small- to medium-size firms (SMEs) were the "right size". SMEs have volume requirements that align with informal sellers' production capacity, and they do not require suppliers to register on complex, standardized platforms. More importantly, they buy "expensively". Consider, for example, a SME that needs 200 pairs of trousers, each with the SME's logo on it. The order calls for 90 pairs of trousers for women and 110 pairs for men. Of these, 45 need to be small size, 52 medium, 63 large, and 40 extra-large. Big producers are rarely interested in accepting

[7] https://www.JUNTOScomprasconimpactosocial.com/.

such an order. SMEs therefore end up buying from suppliers further down the distribution chain because usually only small retailers will accept such specific orders. Thus, buying from a cooperative is often cheaper.

We then turned to the question of *how JUNTOS should operate*. We decided that as a permanent new market actor, JUNTOS should operate as a marketplace. However, regarding operations, there were several issues that needed to be addressed. First, should JUNTOS be a virtual platform? The main LATAM virtual marketplace, Mercado Libre, helped us answer this question.[8] Mercado Libre launched as a social marketplace. They provide informal sellers with support, which includes sending their professionals to teach them how to use the platform and produce quality photographs of their products for posting on it. Our JUNTOS team held a series of meetings with the person responsible for this project within Mercado Libre. She explained that the cooperatives rarely have the skills to upload photos of their products. Also, because all members of these cooperatives are busy making products, typically by hand, they usually only look at their cellphones late at night, thus losing many sales.

These insights were validated during our face-to-face visits to the cooperatives. As a person in one of them explained, "When assisted by Mercado Libre, I was amongst the first five options in a given search. When they stopped supporting me, I dropped to after page ten, so I did not receive any new orders. I must be doing something wrong, but I don't know what it is." We also observed that most of the platform buyers for the impact products were end consumers (i.e., business-to-consumer) rather than companies (i.e., business-to-business). With these considerations in mind, we decided JUNTOS should operate as a non-digital marketplace.

We also knew that JUNTOS, as an intermediary, needed to provide informal sellers with access, through bridging social capital, to SME buyers. We did this by identifying second-order organizations that aggregate SMEs. For example, we arranged for JUNTOS to form an alliance with Argentina's National Franchising Association, which clusters all franchisors. Another example was the agreement we reached with an important international bank. As part of their corporate social responsibility program, the bank agreed to advise its SME clients to initiate inclusive purchases, with JUNTOS operating as the actor able to solve their purchasing needs.

Our next consideration concerned the need for JUNTOS to provide the cultural capital that informal sellers lack but require to facilitate exchanges with buyers. Cultural capital takes many forms. For example, we encountered that cooperatives did not keep "regular" nine to five working hours. To keep up with family duties, members of the cooperative often worked in the evenings. As a result, JUNTOS needed to have staff available to SMEs during the SMEs' nine to five schedules as well as staff who could talk to the cooperative in the evenings. The cooperatives also lacked familiarity with using emails. We arranged for all conversations and selling arrangements to take place through WhatsApp. Thus, for example, after receiving through WhatsApp a cooperative's price for an order, JUNTOS could set out the details of the transaction in a formal order form and send it, with appropriate wording,

---

[8] https://mercadolibre.com/.

to the SME customer. We discuss more such examples below when addressing the one-stop-shop challenge. Figure 1.3 shows the new inclusive value chain proposed.

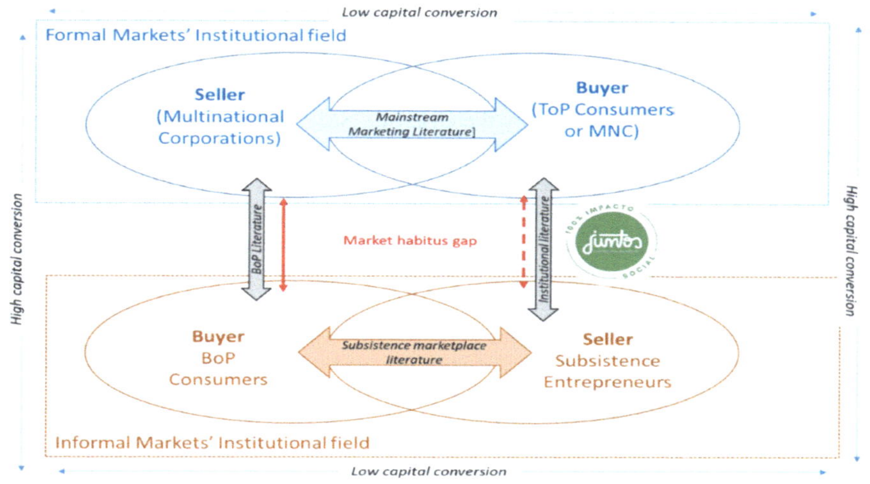

Note: TOP, top of the pyramid; BOP, base of the pyramid; MNC, multinational corporation.

Fig. 1.2 The market habitus gap and the new permanent intermediary (adapted from Pels et al., 2022). (*Note* TOP, top of the pyramid; BOP, base of the pyramid; MNC, multinational corporation)

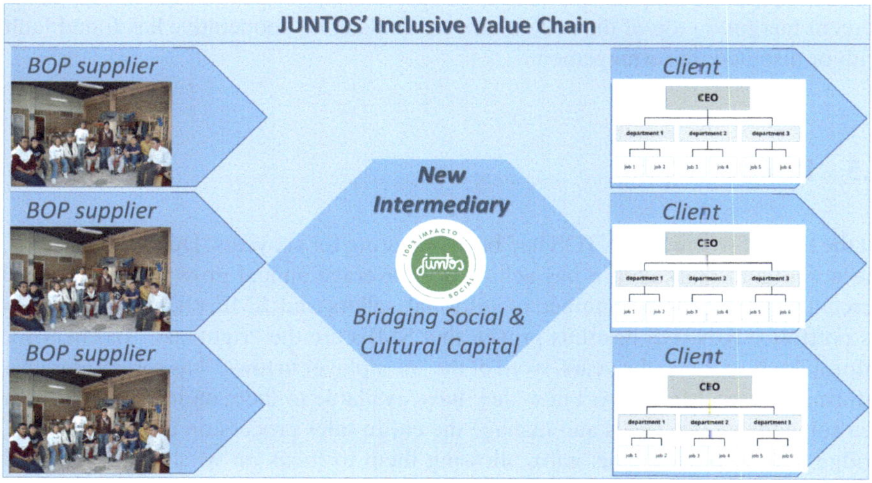

Fig. 1.3 JUNTOS' proposed value chain

The next question we asked ourselves was *how should JUNTOS handle lack of working capital?* We were well aware that even though SME orders are relatively small, informal sellers still need to have working capital to buy raw materials. We accordingly created a policy whereby the SMEs would pay 50 percent of the total price in advance of receiving the order and the remaining 50 percent on delivery of the order. Since then, we have learned that SMEs have no problems with this policy.

Could JUNTOS become a one-stop-shop? One of the challenges to achieving this objective can be exemplified by the case of a SME ordering a welcome kit that includes a pen, notebook, mug, tee-shirt, and a wooden support for a laptop, with all items sporting the SME's logo. An order such as this implies the need to coordinate several informal sellers within the one sales order. Experience showed that if our JUNTOS team worked with each informal seller independently (rather than trying to make the cooperatives coordinate themselves), then the one-stop shop business proposal could be readily achieved.

And what about cross-selling? Was that possible? Many SMEs start out by buying something from an informal seller, for example, end of year gifts. The challenge for JUNTOS at this point is to cross-sell other product categories (produced by other informal sellers) to the SME. Examples might include stationery, work uniforms, catering (food), and the like. We soon found that once SMEs gained trust in JUNTOS, cross-selling was not a problem.

We also wanted to determine *if charging a commission would be problematic.* Because an important part of our action-research work was determining if JUNTOS would be a viable business proposition, we had to identify sources of income. Our JUNTOS team talked to the informal sellers in a way that made fully clear to them the advantages JUNTOS was offering, notably bridging social capital and managerial know-how. Eventually, we reached agreements with the informal sellers whereby the prices they put forward would not be changed but JUNTOS would add a 10 to 35 percent margin on top of the sellers' prices. To date, no cooperative has found fault with or disputed this arrangement.

1.5 Results

At the time of writing, JUNTOS had been operating for six years. During this period, the non-digital marketplace has achieved an average annual growth of above 100 percent in terms of the number of informal sellers and SME clients included in its portfolio. JUNTOS has thus proven that SMEs are the "right size" of client for informal sellers. Over the years, word of the "new player in town" has spread amongst informal sellers. Many now know they have available to them an intermediary that can solve their sales needs and manage the entire sales process on their behalf (i.e., bridge their market habitus gap), allowing them to focus on what they know and do best—their craft and skills. We have also seen an increasing number of informal sellers taking the initiative in terms of tracking down JUNTOS and asking to join its pool of suppliers. Currently, JUNTOS's operation encompasses eight categories of product: textiles (e.g., uniforms), stationery, carpentry, catering, gifts, industrial hygiene, and printing. JUNTOS has also been able to de-seasonalize sales through cross-selling and successfully coordinate up to five informal sellers in a single one-stop-shop order. To date, neither the 50 percent advance payment nor the commission has posed problems.

JUNTOS's success has generated diverse spillovers. It has developed alliances with various second-order organizations (e.g., the National Franchising Association) that want to offer their associates the possibility of purchasing their non-strategic goods and, in so doing, generate a social impact. Additionally, in 2022, the government of the City of Buenos Aires (DGAPP) hired JUNTOS to transfer knowledge on how to evaluate their pool of informal sellers as well as to include them in JUNTOS' sales catalogues and to train four organizations to replicate the JUNTOS model. In 2024, the DGAPP extended the knowledge transfer contract.

Because JUNTOS's development is the outcome of our action-research project, we have kept records of all research procedures and learnings. We have compiled this information into a JUNTOS Guide that allows to replicate JUNTOS and can be downloaded for free from the ENI website.[9] Recently, we signed a memorandum of agreement with ProMujer, a regional social enterprise that has been working to advance gender equality in Latin America for more than 33 years.[10] ProMujer's Gender Knowledge Lab, the consulting arm of ProMujer, specializes in providing solutions that support both public and private institutions as well as NGOs, is interested in adopting and adapting the advice in the JUNTOS Guide.

1.6 Contributions and Implications

The pilot project, the in-depth literature review (Pels et al., 2022), JUNTOS's six years of development and growth, and the contracts the JUNTOS has established with the DGAPP have provided our action-research team with many insights. The literature review allowed us to understand why the pilot project fared so poorly, specifically because the problem it sought to resolve had been defined as one of *transitioning between* markets (i.e., a regulative type of problem/solution). We suggested an alternative solution, one that recognizes the need to bridge the market habitus gap and thereby position the problem and solution as *access to* formal markets (i.e., a cultural-cognitive type of problem/solution). Finally, we decided a new market actor was needed, namely, a permanent intermediary that would enable a more inclusive purchasing solution.

Our development of JUNTOS also allowed us to claim the existence of a market void. While market voids have similarities with institutional voids, they also differ from them. Mair and Marti (2009) adopted the term institutional void "to describe a situation where absent and/or weak institutional arrangements prevent those excluded

[9] https://www.utdt.edu/ver_contenido.php?id_contenido=11125&id_item_menu=21531.

[10] https://promujer.org/portal/.

by poverty from participating in market activities" (p. 2). The pilot project as well as the JUNTOS case shows that there is a void but it isn't institutional. Thus, a definition of a market void is put forward. A market void denotes *a situation in which a market actor is missing and whose role cannot be fulfilled by current market players, thus impeding market exchange.*

The JUNTOS action-research case also validated the insights suggested by Pels et al (2022). The action-research revealed deep cultural-cognitive differences and confirmed that because different logics underpin formal and informal marketplaces (i.e., different habitus), informal sellers have difficulty acting in formal markets. This understanding again convinced us, at the ENI, of the need for a new permanent actor.

The findings of our ongoing JUNTOS action-research provided us with additional insights. These emphasized that the solution should not be a digital marketplace because informal sellers generally produce their products by hand and have very limited digital skills. We agreed that a more traditional form of interaction between these sellers and their buyers/customers would work best. We also concluded that the ideal customer would be SMEs rather than big MNCs with sustainability departments. Furthermore, JUNTOS provided evidence that second-order organizations are willing to become allies of one-stop-shop providers of inclusive purchasing solutions. We also found that informal sellers' lack of managerial skills extends beyond difficulty preparing formal purchase orders to the difficulty of coordinating with other cooperatives, a situation that can mean loss of potential sales. As an intermediary, JUNTOS allowed us to see that one-stop-shop solutions represent the majority of SME orders. For example, an SME might order an end of year gift box containing catering products from a cooperative but also want some art or craft items or gifts for small children only available from another cooperative. We soon found that SMEs, after a first positive purchasing experience, are willing to cross-purchase into other categories, thus increasing the bridging social capital that JUNTOS, as an intermediary, can provide. Finally, the JUNTOS action-research findings negated the belief that charging a commission in an inclusive value proposition impedes sales.

An overarching theme is the dialogue between the conceptual and empirical discussions presented in this chapter and their alignment with the Sustainable Development Goals mentioned at the beginning of this chapter. Two of these goals relate to the chapter's core ideas—SDG 9 (Industry, Innovation, and Infrastructure), and SDG 17 (Partnerships for the Goals). SDG 9 focuses on inclusive, sustainable industrialization and fostering innovation. It calls for investment in infrastructure and innovation directed toward providing small businesses and marginalized communities with better access to wider markets and commensurate opportunity to scale their operations sustainably. The novel permanent intermediary solution (in our case, JUNTOS) contributes to SDG 9 because its one-stop-shop and cross-selling policies make it possible to scale the informal sellers' operations. Moreover, by providing both social bridging capital and cultural capital, it enables BOP sellers to overcome their market habitus gap. It levels the "playing field" for BOP sellers by reducing disparities and enhancing the sellers' value propositions and access to the formal sellers (SMEs), thus facilitating economic growth. SDG 17 emphasizes the importance of multi-stakeholder partnerships. It maintains that collaboration between governments, businesses, and civil society can create pathways for small producers to access markets, thereby enhancing their ability to sell products and contribute to sustainable development. The JUNTOS case also helps confirm that second-order organizations along with other social enterprises and governmental institutions are willing to collaborate and see value in doing so.

In terms of public policy implications, the JUNTOS action-research project's interactions with governmental organizations reveal that there is often a strong regulatory approach to identifying problems and solutions. Large public resources, both human and financial, are often invested in programs supporting BOP sellers, which typically offer various combinations of mentoring, training, and microfinancing. However, once sellers complete these programs, they are often "left on their own," as resources are then directed toward new groups of beneficiaries. Efforts to secure potential customers for BOP sellers generally accompany these programs; for example, through government-sponsored inclusive fairs. However, these events do not include intermediation support. Informal sellers, participating in these fairs, have reported achieving only limited commercial success. Our experience with JUNTOS's consultancy programs suggests that governmental programs might want to shift from training programs to the creation of a permanent intermediary actor that can help coordinate cross-selling and one-stop purchasing situations.

Overall, JUNTOS's effectiveness invites a call to action. Whereas Yunus (1999) identified that microcredit institutions were missing, the findings of the JUNTOS action-research strongly indicate that that another type of intermediary were missing. We argue that the notion of market void invites us to consider what other market actors are absent, preventing informal and formal markets from interacting to their mutual benefit.

References

Avison, D. E., Lau, F., Myers, M. D., & Nielsen, P. A. (1999). Action research. *Communications of the ACM, 42*(1), 94–97. https://doi.org/10.1145/291469.291479

Banerjee, A. V. (2013). Microcredit under the microscope: What have we learned in the past two decades, and what do we need to know? *Annual Reviews of Economics, 5*(1), 487–519.

Bourdieu, P. (1977). *Outline of a theory of practice*. Cambridge University Press.

Bourdieu, P. (1986). The forms of capital. In J. G. Richardson (Ed.), *Handbook of theory of research for the sociology of education*.

Bourdieu, P. (1999). *Distinction: A social critique of the judgement of taste*. Routledge.

Bourdieu, P. (2005). *The social structures of the economy*. Polity.

Bradley, S. W., McMullen, J. S., Atmadja, A., Simiyu, E., & Artz, K. (2011). Self-employed or employing others? Pre-entry capabilities, entrepreneurial action, and the learned resourcefulness of microcredit firm founders. *Frontiers of Entrepreneurship Research, 31*(19), 643–658.

Brundtland, G. H. (1987). *Our common future: Report of the World Commission on Environment and Development*. UN-Dokument A/42/427.

Bruton, G. D., Khavul, S., & Chavez, H. (2011). Microlending in emerging economies: Building a new line of inquiry from the ground up. *Journal of International Business Studies, 42*(5), 718–739. https://doi.org/10.1057/jibs.2010.58

DiMaggio, P. (1988). Interest and agency in institutional theory. In L. G. Zucker (Ed.), *Institutional patterns and organizations: Culture and environment* (pp. 3–21). Ballinger.

Dutt, N., Hawn, O., Vidal, E., Chatterji, A., McGahan, A., & Mitchell, W. (2016). How open system intermediaries address institutional failures: The case of business incubators in emerging-market countries. *Academy of Management Journal, 59*(3), 818–840. https://doi.org/10.5465/amj.2012.0463

Gehman, J., Glaser, V. L., Eisenhardt, K. M., Gioia, D., Langley, A., & Corley, K. G. (2018). Finding theory–method fit: A comparison of three qualitative approaches to theory building. *Journal of Management Inquiry, 27*(3), 284–300. https://doi.org/10.1177/1056492617706029

Gittell, R., & Vidal, A. (1998). *Community organizing: Building social capital as a development strategy*. Sage Publications.

Granovetter, M. (1977). The strength of weak ties. *Social Networks, 78*, 347–367. https://doi.org/10.1016/B978-0-12-442450-0.50025-0

Granovetter, M. (1985). Economic action and social structure: The problem of embeddedness. *American Journal of Sociology, 91*(3), 481–510. https://faculty.washington.edu/matsueda/courses/590/Readings/Granovetter%20Embeddedness%20AJS.pdf

International Labour Organization. (2023). *Women and men in the informal economy: A statistical update*. https://www.ilo.org/publications/women-and-men-informal-economy-statistical-update

Karlan, D., & Valdivia, M. (2011). Teaching entrepreneurship: Impact of business training on microfinance clients and institutions. *Review of Economics and Statistics, 93*(2), 510–527. https://doi.org/10.1162/REST_a_00074

Kistruck, G. M., Beamish, P. W., Qureshi, I., & Sutter, C. J. (2013). Social intermediation in base-of-the-pyramid markets. *Journal of Management Studies, 50*(1), 31–66. https://doi.org/10.1111/j.1467-6486.2012.01076.x

Little, P. D., & Dolan, C. S. (2000). Adjustment in Sub-Saharan Africa. In A. Haugerud, P. Stone, & P. D. Little (Eds.), *Commodities and globalization: Anthropological perspectives* (pp. 59–78). Rowman and Littlefield.

Loayza, N. (2018). *Informality: Why is it so widespread and how can it be reduced?* World Bank Research & Policy Briefs, No. 20. World Bank. http://documents.worldbank.org/curated/en/130391545228882358/Informality-Why-Is-It-So-Widespread-and-How-Can-It-Be-Reduced

London, T., Anupindi, R., & Sheth, S. (2010). Creating mutual value: Lessons learned from ventures serving base of the pyramid producers. *Journal of Business Research, 63*(6), 582–594. https://doi.org/10.1016/j.jbusres.2009.04.025

Mair, J., & Marti, I. (2009). Entrepreneurship in and around institutional voids: A case study from Bangladesh. *Journal of Business Venturing, 24*(5), 419–435. https://doi.org/10.1016/j.jbusvent.2008.04.006

McMullen, J. S. (2011). Delineating the domain of development entrepreneurship: A market-based approach to facilitating inclusive economic growth. *Entrepreneurship Theory and Practice, 35*(1), 185–215. https://doi.org/10.1111/j.1540-6520.2010.0042

North, D. C. (1990). *Institutions, institutional change, and economic performance*. Cambridge University Press.

Pels, J., Araujo, L., & Kidd, T. A. (2022). Informal sellers and formal markets: A habitus gap. *Journal of Business & Industrial Marketing, 37*(6), 1269–1280. https://doi.org/10.1108/JBIM-03-2021-0180

Pels, J., & Álvarez Giraldo, Y. (2015). *Programa Generar* [Generar Project]. Espacio de Negocios Inclusivos de la Universidad Torcuato Di Tella.

Prahalad, C. K. (2004). *The fortune at the bottom of the pyramid: Eradicating poverty through profits*. Wharton School Publishing.

Scott, W. R. (2014). *Institutions and organizations: Ideas, interests, and identities* (4th ed.). Sage Publications.

Shakespeare, W. (1603/n.d.). *The tragedy of Hamlet, Prince of Denmark* (B. Mowat, P. Werstine, M. Poston, & R. Niles, Eds.). The Folger Shakespeare. https://www.folger.edu/explore/shakespeares-works/hamlet/read/

Siggelkow, N. (2007). Persuasion with case studies. *Academy of Management Journal, 50*(1), 20–24. https://doi.org/10.5465/AMJ.2007.24160882

Sutter, C., Webb, J., Kistruck, G., Ketchen, D. J., Jr., & Ireland, R. D. (2017). Transitioning entrepreneurs from informal to formal markets. *Journal of Business Venturing, 32*(4), 420–442. https://doi.org/10.1016/j.jbusvent.2017.03.002

United Nations. (n.d.). *Sustainable development goals: 17 goals to transform our world*. https://www.un.org/en/exhibits/page/sdgs-17-goals-transform-world

Universidad Torcuato di Tella. (n.d.). *Eni di Tella*. https://www.utdt.edu/ver_contenido.php?id_contenido=11120&id_item_menu=21522

Viswanathan, M., Sridharan, S., Gau, R., & Ritchie, R. (2009). Designing marketplace literacy education in resource-constrained contexts: Implications for public policy and marketing. *Journal of Public Policy & Marketing, 28*(1), 85–94. https://doi.org/10.1509/jppm.28.1.85

Webb, J. W., Kistruck, G. M., Ireland, R. D., & Ketchen, D. J., Jr. (2010). The entrepreneurship process in base of the pyramid markets: The case of multinational enterprise/nongovernment organization alliances. *Entrepreneurship Theory and Practice, 34*(3), 555–581. https://doi.org/10.1111/j.1540-6520.2009.00349.x

Woolcock, M. (2001). The place of social capital in understanding social and economic outcomes. *Canadian Journal of Policy Research, 2*(1), 11–17.

Woolcock, M., & Narayan, D. (2000). Social capital: Implications for development theory, research, and policy. *The World Bank Research Observer, 15*(2), 225–249. https://doi.org/10.1093/wbro/15.2.225

Yunus, M. (1999). *Banker to the poor: Micro-lending and the battle against world poverty*. Public Affairs.

Jaqueline Pels is Full Professor of Marketing at the University Torcuato Di Tella Business School, Buenos Aires, Argentina. She is Director of the Inclusive Business Think Tank (ENI-DiTella). Her research interest is in the areas of inclusive business, emerging economies, marketing theory, relationship, and networking marketing, as well as the Service Dominant logic. Her publications have appeared in leading international journals, including Journal of the Academy of Marketing Science, Marketing Theory, Journal of Business Research, Journal of Business and Industrial Marketing, European Journal of Marketing, Journal of Relationship Marketing, amongst others. She has served on the Editorial Boards of the Journal of Marketing, Marketing Theory, Journal of Business and Industrial Marketing, amongst others.

Chapter 2
Inclusive Marketing: Fostering Sustainable Wellbeing in Subsistence Communities

Srinivas Venugopal and Anaka Aiyar

Abstract While marketers have been immensely successful in catering to the needs of consumers in contexts of affluence, the ultimate challenge confronting marketers is to cater to the underserved needs of low-income consumers around the globe. Ensuring that individuals from low-income contexts have access to quality products and services that can improve their welfare is the greatest opportunity of inclusive marketing. If we do not take inclusive marketing seriously, we run the risk of sustaining unjust societal inequities, or worse, we become complicit in accentuating them. In this chapter, we propose ideas to create an inclusive marketing mindset that address unserved needs (a) by creating new marketplace exchanges where none existed or (b) by supplanting a less optimal marketplace exchange with one that balances the process of creating social, environmental, and economic value. To do so, we highlight 8 mistakes commonly made by marketers who seek to improve the well-being among low-income communities in a sustainable manner. With examples and reflections on our research and practice, we propose ideas for what an inclusive marketer can do to avoid these mistakes. We also propose ways to make marketing practices inclusive and sustainable in addressing underserved needs in various low resource contexts.

Keywords Social innovation · Subsistence · Base-of-the-pyramid · Poverty · Development

---

S. Venugopal (✉) · A. Aiyar
University of Vermont, Burlington, VT, US
e-mail: Srinivas.Venugopal@uvm.edu

A. Aiyar
e-mail: Anaka.Aiyar@uvm.edu

R. Singh and T. A. Wani (eds.), *Marketing to the Base of the Pyramid*,
https://doi.org/10.1007/978-981-96-7105-2_2

2.1 The Promise of Inclusive Marketing

Most days, on the way to work, we purchase a cup of cappuccino for about 4 dollars and do not even think twice about it. It is staggering to think that there are over 2 billion people globally who live on less than 4 dollars a day (World Bank, 2024). Much like you and me, these two billion people have needs such as nutrition, education, credit, healthcare, and recreation that are critical to their wellbeing. But unlike you and me, they must meet these needs with less than 4 dollars a day. While marketers have been immensely successful in catering to the needs of consumers in contexts of affluence, the ultimate challenge confronting marketers is to cater to the underserved needs of low-income consumers (Singh, 2015). Ensuring that individuals from low-income contexts have access to quality products and services that can improve their welfare is the greatest opportunity for inclusive marketing. If we do not take inclusive marketing seriously, we run the risk of sustaining unjust societal inequities, or worse, we become complicit in accentuating them.

Inclusive marketers should set a high aspirational bar for themselves. Not only should they serve the needs of the poor in an economically sustainable manner, but they should do so while rejuvenating the environment and adding to the vitality of local communities (Pingali et al., 2019). This vision requires marketers to rethink the way they design and develop products, services and processes and requires them to reimagine business models to be more inclusive and sustainable for the people, planet, going beyond profits. In other words, inclusive marketing necessarily entails an entrepreneurial mindset that creates a new world that allows people, the planet, and value to be jointly determined (Venugopal, 2016). Low-income markets pose unique challenges for marketers in creating mutually beneficial marketplace exchanges (Viswanathan & Rosa, 2010).

In this chapter, we propose ideas to create an entrepreneurial mindset that addresses unserved needs (a) by creating new marketplace exchanges where none existed or (b) by supplanting a less optimal marketplace exchange with one that balances the process of creating social, environmental, and economic value. The process of inclusive marketing, when optimally orchestrated, can bring about marketplace inclusion of underserved communities who are subject to multiple forces of marketplace exclusion (Saatcioglu & Ozanne, 2013). The challenge of using inclusive marketing to address unmet needs in contexts of poverty has been the central force that has animated our professional life. Prior to moving to academia, the authors of this chapter held leadership roles in a technology-based social venture at the Indian Institute of Technology Madras, whose mission was to provide wellbeing enhancing products and services to consumers in far flung and sparsely populated villages of South India. India has over 600,000 villages and a typical village has a median population of less than 800 people. We learned in the course of running our ventures that critical material resources were concentrated in urban centres that were far removed from India's villages. The online mode of service delivery, a novel idea in the mid-2000s when we were working on the field, provided one viable solution to address the gap of resource constraints. Leading these ventures provided us with a ringside view of the transformative impact that marketers can have in the lives of people if they could be focused on catering to well-being enhancing needs such as providing education. For many of our consumers, online services opened up the opportunity to pursue higher education and enhance their household incomes. These pathways of upward socio-economic mobility are often denied to resource constrained consumers in rural markets.

2.2 Common Mistakes that Thwart Inclusive Marketing Efforts

During our engaged research in BoP settings, we witnessed some recurring mistakes on the part of inclusive marketers. In this chapter, we will walk the reader through some of these mistakes and then discuss our views about how one can avoid these pitfalls. This chapter is interspersed with exhibits of mini cases of inclusive marketing efforts addressing needs in underserved communities across diverse geographies and times. These case studies are not meant to be presented as gold standards on inclusive marketing practices. Instead, we caution the reader to analyze these with a critical eye to obtain a deeper understanding of the promise and pitfalls when engaging in ideas for inclusive marketing. Our chapter underscores the importance of deriving insights from human stories that often undergird abstract scientific theories such as poverty traps or inform cold statistics about the nature of poverty. Therefore, each of the following sections will begin with an orienting human story that alerts us to core problems that plague inclusive marketing efforts. We will then provide reflections from our research and experience as practitioners on how to avoid these mistakes.

2.3 Mistake 1: Rich Man's View of a Poor Man's Needs

The first mistake we would like to discuss is what we call "Rich man's view of a poor man's needs." Inclusive marketers spring into action whenever they detect unfulfilled needs in the markets. Typically, most marketers do not have lived experience of poverty themselves. Therefore, they enter contexts of poverty with a top-down view that is ill-suited to developing cogent solutions. A story from our fieldwork illustrates this point. John is a young adult, who lives in the refugee camp in Uganda. During Srini's interaction with him, John told him that he considers himself lucky if he gets one square meal a day. His typical day included getting up early in the morning, drinking a lot of water, and then playing soccer all day long. John ate his one square meal a day in the night before going to bed. As outsiders, we found John's way of organizing his day puzzling. If someone is calorie starved, why would they spend the entire day playing soccer, which requires intense physical activity? And, if they could only eat one square meal a day, why would they not eat it in the morning (so that it sustains one throughout the day)? John explained to Srini that the reason he played soccer all day long was because it distracted him from the hunger that he felt. The reason he ate his one square meal a day at night was because he could go to sleep if he was hungry. This was a very humbling experience for us early in our academic career. As outsiders, as an "expert" on nutrition, our theory-based understanding of John's needs completely failed to grasp his relived reality of poverty. While we were focused on thinking about helping John get more "nutrition," John was interested in solving his problem of "hunger."

A key lesson for us and the inclusive marketer would then be that when one is trying to create well-being enhancing solutions for low-income settings, it is crucial to start with an understanding of the lived realities of people in these settings, as well as the contextual realities. Determining the needs of the local community requires a humility, openness, and connection to members for whom we want to envision solutions. This process also involves reflection that we are not imposing our pre-existing knowledge and to identify and unlearn commonly held beliefs or misconceptions of the people we work for before immersing oneself in similar low-income contexts. The process of unlearning and reframing our understanding of what are key needs is the first step towards designing appropriate solutions that can address underserved needs.

2.4 Mistake 2: Designing Solutions for the "Other"

In creating solutions for low-income markets there is a serious trap of "othering." Othering happens when a naive inclusive marketer proposes to create one type of product for the poor and another type of product for the rich. Here the value created for the product (for the poor) is focused on value generation from the richer consumer instead maximizing value for the poor consumer. Let us illustrate through an example.

Exhibit 1: Playpump—Solving needs through fun activities

The picture above shows a drinking water solution called PlayPump that generated millions of dollars of funding. PlayPump was billed as a magic-bullet solution to Africa's water problems. The idea of PlayPump was that children playing in a merry-go-round in their school would pump groundwater into an overhead tank that can then dispense water for the school or community. The overhead tank could be used for advertising thereby generating income for the school or community. The solution seemed sensational on paper and generated a lot of excitement and support from actors such as Jay-Z and the Case Foundation. But very soon, PlayPumps installed across villages started failing one by one. Why?

First, can one imagine the children of Wall Street executives spinning a merry-go-round to get water into their homes? Yet, investors were more than happy to convince themselves that the "poor" would find value in such a product. Second, the solution was clearly developed without much consultation with community members since there was no operational support once these towers were built. This meant that the towers, once installed, could not be fixed or maintained. Third, if one took a closer look at the language of the advertisement in the PlayPump picture, one would realize that the advertising is in English. Most villagers in Mozambique cannot read, let alone read English. Hence the only takeaway here was that the product was designed to elicit the interest of outside elites to raise funding for the cause. If communities did find value in such a product, it was not evident in their eventual dysfunction. Solutions that start by claiming to design products exclusively for the poor are often driven by stereotypes of the poor and often exacerbate the problem instead of addressing them.

Hence, an inclusive marketer who has engaged with their community will design novel products and business models that will be more optimal for all consumers, poor or rich alike. In this chapter, we also propose that being more environmentally and socially sustainable while taking into consideration economic sustainability is the ultimate challenge for the inclusive marketer.

Exhibit 2: Soccket

If you think this is a unique case that is not representative of a more general and recurring malaise, here is another example.

Many villages in Ghana are yet to be electrified, which means that children cannot read or do their homework after sundown. Soccket is a soccer ball that charges a battery when it is kicked around. The designer claims that children can fix an LED light to the soccer ball to read or complete their homework at night. Again, the usual script played out. The solution was hailed as a sensational and creative solution for the poor. No one stopped to ask how they would feel if their own children had to kick around a soccer ball all day so they could read at night. Yet, we are willing to accept that this is a persuasive solution for the "poor".

2.5 Mistake 3: Deficit-Based Thinking

The second mistake that is often made by inclusive marketers is engaging in deficit-based thinking. When entering BoP communities, it is quite easy to fall into the trap of looking at what these communities lack. We will give you an example of this again. Millions of dollars are spent every year creating financial literacy programs for Indigenous communities in Australia. The foundational premise of the funding is that Indigenous communities, whose education levels are lower than the population, lack the financial literacy to use money effectively.

Exhibit 3: "Money is not our culture"

Comments from indigenous community members from Australia about their view on money.

"Money is not our culture—is white man culture."

"It's true—because it was introduced to us... money—it was introduced.."

"We never had money, we never."

"Here aboriginal people never used to worry about money, they just went about hunting, do their thing, sing, learning about culture, teach their children and when children grow they teach their children about culture, they taught us to hunt, all that."

"Had dreaming, never worry about money—until white man brought it in."

(Godinho et al., 2017)

Research conducted with Indigenous communities in Australia revealed that communities considered cash exchanges the "way of the white man." Because there was no concept of personal wealth in this community, exchanges in the past had been non-monetary. All resources were collectively owned, shared, or distributed. Exchanges enabled by money were seen as culture imposed by colonial rulers on communities. Thus, what initially seemed like a problem of financial illiteracy, could also be interpreted as an act of cultural resistance against an alien way of life being imposed from the outside. Without such an insight it would be impossible for us to think of solutions that include reconciliation programs alongside the financial literacy programs. Getting buy-in from the community and making financial literacy programs effective required having conversations about reconciliation (Godinho et al., 2017).

In essence, the inclusive marketer's goal should be to avoid deficit-based thinking and challenge conventional framing of phenomenon that are inherited from books, social media, and such. The inclusive marketer avoids extremes of paternalistic thinking where solutions are predetermined based on assumptions about the community's deficits. It is equally critical for the inclusive marketer to avoid the other extreme—romantic thinking—which celebrates the local and ignores insalubrious aspects within the communities they engage with. Both views will disintegrate when any marketer makes contact with communities and immerses themselves in their ground-level realities. Designing solutions that address both the direct and the invisible cultural need for communities requires building trust with communities and deep engagement to understand how the past, present and future needs for communities are inter-linked. As illustrated by this case, what was thought of as a financial literacy effort should have been designed as a reconciliation effort that addresses past injustices that communities have borne.

Exhibit 4: Integrating Traditional Carpet Weavers With Global Demand
Instead of exclusively focusing on addressing deficits, inclusive marketers should identify cultural strengths in subsistence communities and aim to harness them in their design of inclusive marketing efforts. Jaipur Rugs is a carpet company that typifies such a strength-based approach. Jaipur Rugs works with a vast network of over 40,000 low-income artisans from rural India where the tradition of carpet weaving has endured for well over 400 years. Instead of looking at these artisans as lower educated laborers, Jaipur Rugs views these artisans as entrepreneurs with the wellspring of creative ideas. As part of their "artisan originals" initiative, Jaipur Rugs commissions artisans to create their own designs inspired by their own lives and the world around them. This is in sharp contrast with the conventional strategy in the industry where artisans are treated as low-skilled laborers who merely execute designs given by experts as opposed to exercising their own creativity. By allowing artisans to express their creativity, Jaipur Rugs can command a premium position in the marketplace. This is an example of an inclusive marketing firm that aims to build on the strengths of subsistence communities.

Exhibit 5: Meeting Calorie Versus Cash Needs
Another way this deficit-based thinking emerges is when refugees are given food based on careful calorie calculations conducted by experts. Immersive research with refugees shows that they often sell the rations given to them at subsidized rates in the market. On the surface, this would seem to reinforce the view that refugees misuse donations. But when we look deeper at the issue, we realize that the food that refugees are given based on technical calculations does not consider cultural preferences. In addition to meeting nutritional needs, food is a celebration of cultural heritage and can foster cultural cohesion. By selling their rations, refugees obtain some cash with which they can purchase foods that are more in line with their cultural preferences. What they are seeking is not "indulgence" but agency. Therefore, inclusive markets should create the conditions for agency among consumers.

2.6 Mistake 4: Air Dropping Products into Communities

A fourth mistake of the naïve inclusive marketers is to assume that value is embedded within products and effective products can solve any problem regardless of context. Let us give you an example that illustrates this issue. In Srini's field work conducted with the Maasai in Tanzania in the foothills of Mount Meru, it was determined that this community's key challenge was reducing fluorosis, a condition that leads to bone deformity and discoloration. This problem arose due to geological conditions in that region that increased fluoride content in drinking water. A group of environmental engineers created bone char filters to filter fluoride from the water. These bone char filters could also easily be made with locally available materials, thus solving the concern about using local materials in these products. A year after these filters were distributed, tests revealed that the quality of the water going into the filter was better than the quality of the water coming out (in terms of pathogen content). Needless to say, these test results were puzzling. When engineers explored the issue, they found that the lifespan of the bone char filter was about six months. However, most people in the local communities did not change the filter after six months. Typically, in settings with well-developed market ecosystems, there would be local entrepreneurs who could knock on the doors of their customers every six months and provide a new filter, maybe even for a fee. The existence of such entrepreneurial ecosystems is vital for consumers to derive value from a product in a sustainable fashion. However, merely air dropping a solution like a water filter into the setting without thinking about how it enmeshes with the local entrepreneurial ecosystem prevented people from obtaining value from products over the long term.

Therefore, inclusive marketers should pay attention not just to product design, they should also pay attention to the entrepreneurial ecosystem that exists locally. A robust local entrepreneurial ecosystem will help communities to derive and extend the value created from products designed for well-functioning markets. In South America, sub-Saharan Africa as well as South Asia, we have observed the existence of subsistence entrepreneurs who operate survivalist enterprises in the informal economy. It is particularly important for inclusive marketers to determine a mutually beneficial way of collaborating with these entrepreneurs (Venugopal & Viswanathan, 2019). The informal economy accounts for anywhere between 40 and 60% of the entire GDP in parts of the world like sub-Saharan Africa and South Asia (Webb et al., 2014). Yet, many marketers tend to ignore such a pervasive phenomenon that can offer critical help in developing sustainable solutions.

2.7 Mistake 5: Being Market-Driven Instead of Market-Driving

We now turn to the mistake of being market driven. When marketing a product like cereals in affluent settings, marketers don't have to worry about whether people eat cereals or not. The primary concern for marketers in such settings is understanding what brand of cereals people buy and how to persuade consumers to switch over to their brand (Sheth, 2011). In other words, in well-resourced markets, the key decision that consumers are making is what brand to buy. In low-income contexts, the key decision that consumers are making is whether to buy or whether to forego consumption altogether (Viswanathan et al., 2009). Therefore, when operating in BoP markets, it becomes important for marketers to create the market conditions necessary for consumption before discussing how to enhance well-being (Aiyar & Venugopal, 2020). In other words, firms should be market-driving.

Let us illustrate this insight with an example. A group of anthropologists who were working in parts of South India found that a lot of pregnant women in these communities did not consume iron rich foods when they were pregnant (Nichter, 2008). This observation was puzzling because iron is extremely important for the proper development of the fetus. The anthropologists tried to examine why this was occurring, and what they found was that there was a cultural preference for smaller babies. The reason being that 30 or 40 years ago, when people didn't have access to hospitals, most deliveries would happen at home. It was assumed that if a child was smaller in size, then there would be fewer completions during delivery. Consequently, local norms emerged around preference for smaller babies. In a setting with such cultural norms, marketers would be ineffective in increasing access to "free iron pills" if they framed the issue as being one of affordability or an "economic" constraint. Cultural norms will have prevented the consumption of this well-being enhancing product even when it was free. Inclusive marketers then realize that market development efforts require them to create conditions necessary for the consumption of such well-being enhancing products.

To do so the inclusive marketer explores all facets of their community, including the socio-cultural institutions that may govern people and their cultural beliefs regarding consumption.

2.8 Mistake 6: Overlooking the Problem of Elite-Capture of Value

The next mistake is ignoring the problem of elite capture in local communities. In work done with a small subsistence fishing community in South India over the last decade, we have come to find that fishers in the community complain that traditional knowledge of fishing has become redundant. This is a fishing village of approximately 300 households. The way of life in the community is such that when a child is 8 years old, they drop out of school and go on the boat with elders in the community to learn traditional wisdom on how to navigate the sea. Elders share generational wisdom on where, what time, and how fishers can expect to find a certain type of fish. Much of early adult life is spent learning traditional wisdom regarding navigating the sea. However, when speaking with community members more recently, many say that traditional knowledge on when and where to fish is no longer accurate. The fishing landscape has changed. In other words, traditional wisdom has completely stopped working in these communities. While communities attribute changes to the Tsunami in 2004, there is also a general recognition that disruptive environmental changes from climate change have played a role. Communities who have relied on traditional knowledge for their livelihoods are now worried about their future as fishers (Venugopal & Chakrabarti, 2022).

A group of social entrepreneurs who were trying to solve this problem approached it through the lens of what is called "shared value creation" (Kramer & Pfitzer, 2016). Economists developed the notion of shared value creation, and it holds that multiple stakeholders can come together to address an important societal issue, thus having a positive impact on everyone. The problem with the shared value creation approach is that it completely loses sight of social hierarchy and the lack of equity within BoP communities.

As an illustration, the social enterprise aimed to create shared value by providing GPS and imaging equipment that fishers could purchase. This technology would help fishers determine where schools of fish they were looking for could be found in the sea. While in theory this would increase catch, it was richer Fisher in the community, with bigger boats who were best able to buy imaging and GPS equipment and boost their catch. Poorer fishers who were unable to pay for the GPS were left worse off since the technology enabled over-fishing, thus reducing their catch. In essence, the solution ended up accentuating the inequity in the community.

A key lesson from this experience is that when aiming to create "shared value" it is also important to think about how the value you're creating is going to be shared within the community. An inclusive marketer asks, *"do historically marginalized groups have a seat at the table? Is the division of the value pie equitable?"* These are important questions that inclusive marketers need to attend to while working with communities (Raghubanshi et al., 2021; Venugopal et al., 2024).

Exhibit 6: Addressing Elite Capture

To avoid the problem of elite capture, there are three kinds of activities we suggest for the inclusive marketer.

1. Engage in relational work. Relational work is a process where the services of local stakeholders are identified and leveraged to form mutually valuable exchanges where none existed before. This is a foundational process that enables the creation of shared value by signaling to all stakeholders that maximizing self-interest alone is not the governing principle of the stakeholder network. A good example of such an approach comes from Essilor. Essilor is a major multinational corporation which

is in the vision care business. They estimate that there are over 2 billion people living in far-flung villages of developing countries who need eyeglasses but cannot afford them because distribution channels don't exist in small villages. The impact of this is not just lower quality of life but loss of livelihoods owing to poor vision. The rich from these villages have the mobility to move to bigger cities to get their eyes tested. This is a privilege that many low-income families from these villages do not have. To address this issue, Essilor is recruiting and training rural youth in small villages to become vision entrepreneurs who can test vision and deliver custom eyeglasses that enhance livelihood and quality of life in BOP markets for low-income families.

2. Engage in inclusion work. Inclusion work involves supporting microprocesses necessary to bring about marketplace inclusion of historically excluded groups. Systemic marketplace exclusion can be experienced based on multiple factors such as class, caste, gender, race, or literacy levels (Cummins et al., 2024), and can depend on the local environment in which a particular firm is operating. A good example of inclusion work comes from Evan Cargo. Evan Cargo partners with firms like Amazon to distribute Amazon's products in remote rural areas. Evan Cargo started recruiting women delivery partners after realizing that gender norms within these communities prevented women from finding employment opportunities. Not only did Evan Cargo offer jobs to women, but they also worked with local police proactively to ensure the safety of their women delivery partners. This was important because challenging power hierarchies often create backlash, and Evan Cargo was aware of this and took proactive measures to neutralize the threats faced by their delivery partners.
3. Engage in equity work. The third process is called equity work. Equity work entails the creation of relationships and routines necessary to ensure that shared value is equitably distributed across all intended stakeholders. In a sense, equity work is about ensuring that the most vulnerable members of the value chain are exposed to the least amount of risk. An example of equity work can be found in the work of Jaipur Rugs. Jaipur Rugs is a carpet company that provides sustainable livelihoods to over 40,000 independent artisans who are dispersed across 600 Indian villages. These low-income artisans use their own creativity to design and beef carpets. Jaipur Rugs acknowledges that these small-scale artisans will not be able to absorb the risk of carpets not getting sold to consumers. So, Jaipur Rugs absorbs the risk on their behalf and pays artisans a steady income so that they can run their family. If the carpets are sold, then Jaipur Rugs bears artisans an additional bonus.

(Venugopal et al., 2024)

2.9 Mistake 7: Ignoring Transformative Impacts of Cross-Domain Interdependencies

The next mistake is the mistake of ignoring cross domain interdependencies. To explain, let's do a thought experiment. How does one solve the problem of high absenteeism in schools in BOP markets? Maybe, you thought about improving the quality of teachers in school. If teachers were good, maybe students would be interested in attending school. Or maybe you thought of providing students with a bus service to get to school, making it easier for students to get to the school. However, in low-income settings, the key reason absenteeism rates might be extremely high might have to do with situations that have nothing to do with education.

For example, elder daughters in a household are tasked with taking care of younger siblings. In a situation where financial constraints are tight, where both parents in the household might have to work, it often falls upon the oldest girl child within the household to take care of all the siblings. In this situation, absenteeism may not be driven by the school's resources but instead by a family constraint. As an inclusive marketer, solving the issue required addressing the problem of childcare. An affordable childcare creche that can take care of smaller children might be more effective in keeping girls in school than providing a school bus or investing in high quality teachers. Solving cross-domain interdependencies might hold the key to realizing value when meeting underserved needs among communities.

Exhibit 7: School Meals For Increased Attendance
Another important example is providing breakfast at school to increase attendance. In most countries around the world students are provided with one meal a day during school time. In developing countries, this meal might be the only meal a low-income child will receive during the day. Also, more obvious is that hungry students are less likely to pay attention in class and consequently do worse in tests. This may demotivate students. Providing a healthy and hearty meal may do more to increase attendance and school performance over donating new textbooks in low-income contexts.

2.10 Mistake 8: Relying on Uni-Dimensional Measures of Impact and Progress

Finally, where marketing firms are trying to create solutions for low-income settings, they're also focused on measuring the impact of such initiatives. And typically, you will find that they focus only on the economic impact of the interventions. Marketers often ignore social, psychological, or environmental transformations that might be taking place.

In Srini's longitudinal research of 25 women entrepreneurs in a low-income community in South India, one of his informants told him she owned a bottled pickle business and made a net profit of $10. Upon some basic calculations, Srini was able to deduce that her input costs were high, and her business was most likely not economically viable. When he told her this, she told him, "*The reason I do this is because this is the first time I have money in my hands that I can spend in any way of my own choosing. I don't have to depend on my husband to buy chocolates for my children. It's changed the way I look at myself and what I think I'm capable of."* The key transformations taking place in her life were not economic, but psychological. Additionally, she had to negotiate her subordination in a strongly patriarchal setting. Her husband was not keen on her becoming financially independent and she did not have the option of rebelling. To bring him on board with her financial venture, she started the bottle pickled business that she could run from inside her home. After a few months, he was able to perceive the value she created and was proud to support her when she sold her bottled pickles in the larger city market. Thus, in addition to her psychological transformation, her actions supported a social transformation of her family life (Venugopal et al., 2015). As a reflection, when thinking about impact measurement, focusing exclusively on economic transformations as the sole measure of success, might lead us to miss how inclusive marketing can transform consumers' lives.

Exhibit 8: How To Measure Impact

There are two common pitfalls to be avoided by the inclusive marketer. The first problem stems from imposing definitions of impact commonly accepted at the top of the income pyramid. The second problem stems from thinking of impact as a state to be reached instead of a process of continuous improvement. To illustrate the first problem, in the case of Jaipur Rugs, all the women artisans work from home-based small-scale looms to weave carpets. A case can be made that if these women work in larger looms in centralized factories, they would be more productive and make greater income. But when you speak to women artisans, most of them say that they don't prefer to work in factories because a home-based loom gives them the ability to balance their household needs with the needs of their business, something they derive immense value from. So, when measuring societal impact, we should take into account what impact means from the perspective of those who are to be impacted. Secondly, when we spoke to women artisans, they shared with me that before working with Jaipur Rugs, they would hesitate to leave the threshold of home because they did not feel confident to engage with the world outside because of the gender norms in these communities. But in a few years of working with Jaipur Rugs, they've acquired the confidence to move around within their village and interact with strangers when seen from the perspective of what these women are able to do now that they were unable to do before, this represents a massive societal impact. So, it's important to think of impact as a process of transformation as opposed to a final destination or an ideal state that should be reached.

(Venugopal etal., 2015)

2.11 Conclusion

In this chapter, we reflect on our research and practice providing the reader with a list of common mistakes that inclusive marketers operating in BoP markets should avoid. Examples from this chapter highlight key underserved needs of vulnerable communities across the globe and the pitfalls of marketers being top-down in their beliefs, views, approach, and actions. Inclusive marketing promises to bring transformation into the lives of communities and their members. But to do so the inclusive marketer should engage with their community, listen deeply, being both a romantic who empathizes but also a critic, who challenges the taken-for-granted. Inclusive markets should not just try to "ease the burden of life," but they should envision a future that "adds color to people's lives." While many inclusive marketers may think that it is vulgar to make money by addressing the needs of the poor, they also fail to understand that robbing poor consumers of their agency by assuming they cannot make decisions that are in their best interest, is also ethically problematic.

References

Aiyar, A., & Venugopal, S. (2020). Addressing the ethical challenge of market inclusion in base-of-the-pyramid markets: A macromarketing approach. *Journal of Business Ethics, 164*(2), 243–260.

Cummins, J., Guo, J., Agarwal, N., Aiyar, A., Jain, V., & Bergmann, A. (2024). Caste differences in child growth: Disentangling endowment and investment effects. *World Development, 179*, Article 106598.

Godinho, V., Venugopal, S., Singh, S., & Russell, R. (2017). When exchange logics collide: Insights from remote indigenous Australia. *Journal of Macromarketing, 37*(2), 153–166.

Kramer, M. R., & Pfitzer, M. W. (2016). The ecosystem of shared value. *Harvard Business Review, 94*(10), 80–89.

Nichter, M. (2008). *Global health: Why cultural perceptions, social representations, and biopolitics matter*. University of Arizona Press.

Pingali, P., Aiyar, A., Abraham, M., & Rahman, A. (2019). *Transforming food systems for a rising India* (p. 368). Springer Nature.

Raghubanshi, G., Venugopal, S., & Saini, G. K. (2021). Fostering inclusive social innovation in subsistence marketplaces through community-level alliances: An institutional work perspective. *Industrial Marketing Management, 97*, 21–34.

Saatcioglu, B., & Ozanne, J. L. (2013). A critical spatial approach to marketplace exclusion and inclusion. *Journal of Public Policy & Marketing, 32*(1_suppl), 32–37.

Sheth, J. N. (2011). Impact of emerging markets on marketing: Rethinking existing perspectives and practices. *Journal of Marketing, 75*(4), 166–182.

Singh, R. (2015). Poor markets: Perspectives from the base of the pyramid. *Decision, 42*, 463–466.

Venugopal, S. (2016). *Entrepreneurial marketing in subsistence marketplaces* (Doctoral dissertation, University of Illinois at Urbana-Champaign).

Venugopal, S., Mohapatra, S., & Aiyar, A. (2024). Crafting climate solutions for nature-dependent subsistence communities: A dignity-centric view. *Journal of Macromarketing*, 02761467241238177.

Venugopal, S., Viswanathan, M., & Jung, K. (2015). Consumption constraints and entrepreneurial intentions in subsistence marketplaces. *Journal of Public Policy & Marketing, 34*(2), 235–251.

Venugopal, S., & Viswanathan, M. (2019). Implementation of social innovations in subsistence marketplaces: A facilitated institutional change process model. *Journal of Product Innovation Management, 36*(6), 800–823.

Venugopal, S., & Chakrabarti, R. (2022). How subsistence communities reconfigure livelihood systems in response to climate change: A coupled-systems perspective. *Journal of Macromarketing, 42*(2), 292–307.

Viswanathan, M., & Rosa, J. A. (2010). Understanding subsistence marketplaces: Toward sustainable consumption and commerce for a better world. *Journal of Business Research, 63*(6), 535–537.

Viswanathan, M., Seth, A., Gau, R., & Chaturvedi, A. (2009). Ingraining product-relevant social good into business processes in subsistence marketplaces: The sustainable market orientation. *Journal of Macromarketing, 29*(4), 406–425.

Webb, J. W., Ireland, R. D., & Ketchen, D. J., Jr. (2014). Toward a greater understanding of entrepreneurship and strategy in the informal economy. *Strategic Entrepreneurship Journal, 8*(1), 1–15.

World Bank. (2024). https://pip.worldbank.org/home. Last Accessed 6th/September/2024.

Srini Venugopal is an Associate Professor and holder of the Donald and Gabrielle McCree Endowed Professorship of Business at the University of Vermont's Grossman School of Business. Srini's research demonstrates how social innovation can serve as a potent tool in providing wellbeing enhancing solutions in contexts of poverty. He has published 20 articles on the subject of social innovation based on field data gathered from diverse contexts of poverty spanning five continents. Srini is an award-winning researcher and teacher. His research has won discipline-wide awards from the American Marketing Association and the Association of Consumer Research. For his teaching efforts, Srini was voted the professor of the year by SI-MBA students for three years in a row. Committed to being an engaged scholar, he continues to lead an education-focused social enterprise called Diya that operates in several low-income neighborhoods in South India.

Anaka Aiyar serves as an Assistant Professor at the University of Vermont's Community Development and Applied Economics Department. Anaka's research examines the social determinants of food insecurity and poor health in resource-constrained settings around the world. Employing quasi-experimental research designs with large-scale datasets; her research provides causal evidence on how government policies can effectively improve health and food security among underserved communities. As part of her research program, Anaka has published 15 papers in reputed academic journals. Her open access book, 'Transforming Food Systems for a Rising India,' has been downloaded by over 190,000 readers worldwide. You can learn more about her research at http://www.anaka-aiyar.com/

Chapter 3
Marketplace Literacy and Sustainable Consumption at the Base of the Pyramid (BoP)

Saravana Jaikumar

Abstract In this chapter, we explore the transformative role of marketplace literacy programs in fostering sustainable consumption and empowering consumers at the Base of the Pyramid (BoP). BoP communities face distinct socio-economic and cognitive challenges, including limited literacy, income constraints, and social pressures, which often lead to suboptimal consumption patterns, such as conspicuous consumption. By equipping consumers with essential knowledge, practical skills, and digital literacy, we argue that marketplace literacy programs can address these challenges and enable BoP consumers to make informed, value-oriented choices. We integrate theoretical and empirical insights to highlight how marketplace literacy aligns with Sustainable Development Goals (SDGs), particularly SDG 1 (No Poverty), SDG 4 (Quality Education), and SDG 12 (Responsible Consumption and Production). Our discussion focuses on cognitive and social inhibitions at the BoP, such as difficulties with abstract thinking and susceptibility to persuasive marketing. We examine how marketplace literacy fosters resilience by building metacognitive and social intelligence skills. Additionally, we address the role of ICT-enabled interventions in bridging information asymmetries, empowering women entrepreneurs, and promoting sustainable digital consumption. In the practical applications, we emphasize the importance of community-centric marketing strategies, financial inclusion initiatives, and collaborative partnerships to support sustainable consumption. We also discuss barriers to ICT adoption, including affordability and infrastructure gaps, and propose actionable solutions to overcome these limitations. Through this chapter, we underscore the potential of marketplace literacy programs to create an inclusive, sustainable marketplace that uplifts BoP consumers and contributes to global sustainability goals.

Keywords Marketplace literacy · Sustainable consumption · Base of the Pyramid (BoP) · Conspicuous consumption · ICT interventions · Social intelligence and metacognition

---

S. Jaikumar (✉)
IIM Calcutta, Kolkata, India
e-mail: saravana@iimcal.ac.in

R. Singh and T. A. Wani (eds.), *Marketing to the Base of the Pyramid*,
https://doi.org/10.1007/978-981-96-7105-2_3

3.1 Introduction

The Base of the Pyramid (BoP) represents the largest yet most economically constrained demographic segment, encompassing low-income communities with limited access to mainstream markets, financial services, and quality education. These consumers, often living on less than $5 a day, face unique challenges that impede their ability to make informed marketplace decisions (Viswanathan et al., 2024). At this economic level, limited literacy and constrained cognitive resources compound the obstacles to achieving informed and sustainable consumption, reinforcing cycles of poverty and constraining potential pathways to empowerment (Mishra et al., 2021). These challenges are also reflected in consumption patterns, with BoP consumers often engaging in behaviors driven by social pressures, such as conspicuous consumption, which can further strain their already limited resources (Jaikumar & Sarin, 2015). To cope with these social pressures and to signal social status, many BoP consumers turn to debt, which adds further financial strain and restricts their capacity to make sustainable and economically sound choices (Jaikumar & Sharma, 2021). Given the pressing need to address these interconnected issues, marketplace literacy has emerged as a transformative concept for empowering BoP consumers. Marketplace literacy not only builds consumers' knowledge about the products and services they use but also equips them with critical thinking skills that are essential for navigating complex and often unstructured marketplaces (Viswanathan et al., 2024). For instance, during India's demonetization in 2016, BoP consumers faced significant challenges in navigating the financial ecosystem, highlighting the critical need for interventions such as marketplace literacy programs to enhance their resilience and adaptability.

The objective of this chapter is to examine the potential of marketplace literacy programs as a tool for fostering sustainable consumption among BoP consumers. Marketplace literacy is a holistic approach to consumer education that goes beyond mere functional knowledge, helping consumers understand their rights, identify value, and make decisions based on informed, value-oriented criteria (Jaikumar & Sharma, 2021). This approach is particularly relevant in BoP contexts, where consumers may be more susceptible to exploitative marketing practices and constrained by socio-economic limitations (Viswanathan et al., 2021). As such, marketplace literacy can play a crucial role in shifting BoP consumption patterns from short-term, status-driven purchases to more sustainable, need-based ones (Jaikumar & Sarin, 2015). This shift has profound implications, not only for improving the individual well-being of BoP consumers but also for advancing broader social and environmental goals, particularly those aligned with the Sustainable Development Goals (SDGs).

This book chapter brings together significant research conducted by the author and colleagues in the domains of marketplace literacy, cognitive constraints, and sustainable consumption among BoP consumers. Drawing on a range of studies, including empirical research on the impacts of marketplace literacy programs, the cognitive challenges faced by low-income consumers, and the influence of income inequality on conspicuous consumption, this chapter provides a systematic review of key findings and insights. Insights from studies on ICT-based interventions and the role of digital access for BoP consumers highlight that ICT-mediated interactions can significantly enhance consumer education, support social capital formation, and build economic resilience (Jaikumar et al., 2020; Parthiban et al., 2022). Additionally, research into ICT-enabled value creation for BoP producers underscores the importance of sustained engagement in digital education, which can build marketplace literacy and foster economic resilience among low-income groups (Parthiban et al., 2021).

By addressing these themes, this chapter aims to provide a comprehensive understanding of the potential of marketplace literacy to empower BoP consumers. Through marketplace literacy, BoP consumers can develop essential skills that support both their personal economic well-being and contribute to broader goals of poverty reduction and environmental sustainability (Viswanathan et al., 2024). This chapter underscores the need for inclusive marketing strategies and policy interventions that not only cater to the BoP's unique needs but also respect their dignity, enhance their autonomy, and build pathways to a more sustainable future. This chapter emphasizes the potential of marketplace literacy in contributing to the United Nations Sustainable Development Goals (SDGs), particularly SDG 1 (No Poverty), SDG 4 (Quality Education), and SDG 12 (Responsible Consumption and Production). By fostering informed consumption and sustainable practices, marketplace literacy directly aligns with these global objectives, promoting economic empowerment, reducing poverty, and enhancing the quality of life for BoP consumers.

The chapter is structured into three major sections: **Cognitive and Social Inhibitions at the BoP, Current and Desirable Consumption Patterns and Marketplace Literacy Programs and Other Interventions**. The first section explores the socio-economic and cognitive challenges BoP consumers face, including vulnerabilities that shape their marketplace behavior. The second section analyzes current consumption patterns, contrasting conspicuous and sustainable consumption behaviors, and providing actionable recommendations for fostering sustainability at the BoP. The final section discusses marketplace literacy programs and related interventions, emphasizing their role in building consumer competence.

3.2 Theoretical Background: Cognitive and Social Inhibitions at the BoP

3.2.1 Prevalent Vulnerabilities at the BoP: Cognitive Constraints

BoP consumers are often characterized by economic fragility, low literacy levels, and limited access to formal education. These vulnerabilities exacerbate cognitive constraints, including difficulty engaging in abstract thinking and susceptibility to manipulative marketing practices. Such constraints affect their ability to assess long-term benefits and make informed decisions, resulting in a reliance on short-term, status-driven consumption (Mishra et al., 2021).

BoP consumers often face unique cognitive constraints due to their socio-economic environments, which influence their decision-making processes. Cognitive resources like attention, memory, and executive functioning are often affected by chronic scarcity, leading to decision-making patterns that prioritize immediate rewards over long-term benefits (Mishra et al., 2021). Two cognitive challenges are particularly relevant to BoP consumers: difficulties with abstract thinking and a heightened susceptibility to persuasive messaging. Research indicates that BoP consumers frequently rely on concrete, immediate information, making it difficult to engage in abstract reasoning or understand complex, long-term benefits (Viswanathan et al., 2024). This concrete thinking style often manifests in limited marketplace decision-making, where consumers might opt for short-term gratifications over products that offer better long-term value. By emphasizing immediate and easily understandable information, marketplace literacy programs can help bridge this gap, gradually encouraging abstract reasoning in the consumer domain. Field studies confirm that a bottom-up approach, where concrete examples are introduced before abstract concepts, can be particularly effective in fostering abstract thinking in low-literate, low-income consumers (Viswanathan et al., 2024).

Another significant cognitive constraint is the limited capacity for metacognitive awareness among BoP consumers. Metacognition—thinking about one's thinking—is essential for effective decision-making, especially in marketplaces where consumers must balance their needs with limited resources. BoP consumers, however, often struggle with this due to restricted exposure to educational resources that promote reflective thinking. Marketplace literacy programs that incorporate decision-making frameworks and reflective exercises help consumers build this critical skill, empowering them to assess options more thoughtfully and avoid impulsive purchases (Viswanathan et al., 2024). Further, research emphasizes that consumer education for BoP populations should also address health and welfare choices, especially given the critical role of health literacy in decision-making and quality of life (Jaikumar et al., 2020). Studies indicate that BoP consumers with low health literacy are more likely to overlook preventive measures and engage in riskier behaviors, suggesting that effective marketplace literacy should incorporate health literacy to enhance the overall well-being and resilience of BoP communities.

3.2.2 *Social Influences on Consumption Behavior*

Beyond marketplace literacy, social intelligence and metacognition play essential roles in fostering sustainable consumption among BoP consumers. Social intelligence, or the ability to navigate social complexities, becomes particularly relevant in BoP contexts where social pressures often drive conspicuous consumption. Many BoP consumers engage in visible, status-driven purchases to maintain social standing, a behavior closely linked to income inequality and social comparisons (Jaikumar & Sarin, 2015). When BoP consumers perceive a need to signal their status, they are more likely to divert resources toward conspicuous products rather than essentials. By enhancing social intelligence, marketplace literacy programs can help consumers critically evaluate such social pressures, ultimately encouraging more need-based, sustainable consumption choices.

Metacognition also contributes to sustainable consumption by helping BoP consumers manage the cognitive demands of decision-making. When consumers develop awareness of their thought processes, they are better equipped to identify and avoid manipulative marketing tactics that may exploit their economic vulnerability. Studies have shown that marketplace literacy programs designed with a bottom-up approach—introducing consumers to real-life scenarios first, followed by abstract lessons—significantly enhance BoP consumers' ability to make more sustainable decisions (Sharma & Jaikumar, 2023). By strengthening both social intelligence and metacognitive skills, marketplace literacy helps BoP consumers break out of reactive consumption patterns, fostering a mindset geared toward sustainable, value-driven choices.

Additionally, education and social capital are critical components in promoting health-related decision-making within BoP contexts. A study found that education significantly influences health beliefs, especially when moderated by social capital, which is often less accessible in rural areas (Jaikumar et al., 2020). This underscores the importance of integrating health and financial literacy within marketplace education to empower BoP consumers holistically, supporting informed, sustainable consumption decisions that contribute to long-term resilience.

3.2.3 *ICT-Enabled Marketplace Access and Overcoming Structural Barriers*

The transformative potential of ICT in addressing structural barriers for BoP consumers is well-documented. ICT-enabled tools, such as mobile phones and digital platforms, play a pivotal role in bridging the gaps caused by limited physical market access, informational asymmetries, and poor infrastructure. By providing real-time information on prices, product availability, and financial transactions, ICT reduces the dependency on intermediaries and increases transparency in market participation (Parthiban et al., 2020).

For instance, ICT adoption has proven effective in empowering women entrepreneurs at the BoP, helping them build economic resilience through improved social capital and digital connectivity (Parthiban et al., 2022). Additionally, initiatives such as community-based digital literacy programs have demonstrated significant success in teaching BoP consumers to navigate online marketplaces responsibly. These programs equip users with the knowledge to evaluate product quality, compare prices, and avoid exploitative practices.

Despite its potential, ICT adoption faces critical barriers, including affordability of devices, poor digital infrastructure in rural areas, and gaps in digital literacy. Addressing these challenges requires targeted interventions such as subsidizing low-cost devices, investing in rural broadband infrastructure, and implementing community-based digital training programs (Parthiban et al., 2021). Public–private partnerships, such as government-supported ICT hubs or NGO-led training initiatives, can further enhance the reach and impact of these efforts.

3.3 Conspicuous vs. Sustainable Consumption in BoP Markets

3.3.1 The Psychology of Conspicuous Consumption at the BoP

Conspicuous consumption—the practice of purchasing visible or status-enhancing goods—is common among low-income consumers, driven largely by social pressures and the desire to signal social standing. At the BoP, consumers frequently face high levels of income inequality, which intensifies the urge to signal status through consumption choices (Jaikumar & Sarin, 2015). In this context, material goods often serve as proxies for social standing, particularly in environments where conventional markers of status, such as education or professional achievements, are less accessible (Jaikumar & Sarin, 2015). This behavior can result in a "keeping up with the Joneses" effect, where BoP consumers prioritize visible products over more essential, need-based purchases.

Empirical studies reinforce this tendency in emerging economies. Findings indicate that income inequality within communities' correlates with a higher allocation of resources toward visible consumption items rather than necessities (Jaikumar et al., 2018). Conspicuous purchases may serve as a coping mechanism for perceived social and economic disadvantages, reinforcing both individual and collective status aspirations. However, such consumption patterns often deplete already limited resources, reducing opportunities for upward mobility and financial security. Conspicuous consumption at the BoP is often a response to social comparison pressures, where consumers prioritize visible goods to signal social standing (Jaikumar et al., 2018). However, not all conspicuous consumption is unsustainable. For example, eco-friendly status symbols, such as solar lamps or bicycles, serve dual purposes of status signaling and environmental sustainability. Understanding the cultural norms that dictate such consumption patterns is essential for promoting sustainable alternatives that align with BoP aspirations.

Additionally, digital access through smartphones has been shown to amplify social comparisons within BoP communities. Research reveals that smartphone ownership often becomes a visible indicator of status, further driving consumption of digital goods and services as a means of social signaling (Parthiban et al., 2022). As digital access increases, it may inadvertently heighten the allure of conspicuous consumption, suggesting a need for guidance on sustainable digital consumption within marketplace literacy programs.

3.3.2 The Challenges of Sustainable Consumption for BoP Consumers

While conspicuous consumption offers temporary social rewards, it poses a significant barrier to sustainable consumption among BoP consumers. Sustainable consumption involves making choices that meet immediate needs without compromising future well-being. For BoP consumers, shifting from conspicuous to sustainable consumption requires a focus on need-based products that provide long-term value. However, the cognitive and social pressures influencing BoP consumers' decision-making often hinder this shift. Limited access to information and cognitive constraints related to low literacy make it challenging for BoP consumers to evaluate the long-term benefits of sustainable practices effectively (Viswanathan et al., 2024).

Social dynamics further complicate this shift. In BoP communities, conspicuous consumption often serves as a quick route to social recognition, making it difficult for individuals to prioritize less visible but more sustainable options. These consumption patterns are not only economically restrictive but also environmentally unsustainable, as they tend to prioritize disposable, non-essential goods. Marketplace literacy programs, therefore, play a critical role in reorienting consumer behavior to prioritize sustainability over status-driven purchases (Viswanathan et al., 2021).

Moreover, digital access has potential downsides, as it can expose BoP consumers to persuasive advertising, intensifying desires for non-essential items (Parthiban et al., 2020). By incorporating strategies that highlight sustainable digital consumption, marketplace literacy programs can help mitigate these influences, guiding BoP consumers toward more resilient and value-driven purchasing choices.

3.3.3 Income Inequality and the Implications for Sustainable Consumption

Income inequality plays a significant role in shaping consumption choices among BoP consumers, particularly influencing the balance between conspicuous and sustainable consumption. The stark income disparities within BoP communities can exacerbate status-seeking behaviors, driving consumers to allocate resources toward non-essential, visible goods rather than sustainable alternatives (Jaikumar et al., 2018). As income inequality intensifies, the social appeal of status symbols grows stronger, leading to increased spending on conspicuous items at the expense of long-term needs.

Marketplace literacy programs can mitigate some of the effects of income inequality by providing BoP consumers with tools to identify and prioritize sustainable consumption. By fostering an understanding of concepts like delayed gratification and value accumulation, these programs equip consumers to resist the allure of conspicuous consumption despite socio-economic pressures. As consumers become more educated about the long-term benefits of sustainable choices, they are better equipped to navigate the economic and social complexities of income inequality with resilience. Such knowledge also creates opportunities for sustainable economic growth within BoP communities, as consumers gradually shift from reactive, status-driven spending to proactive, value-driven consumption (Mishra et al., 2021).

Moreover, the integration of ICT-enabled marketplace literacy can help bridge gaps caused by income inequality, providing BoP consumers with a broader range of sustainable choices. Studies have shown that digital access not only enables better market participation but also helps low-income consumers access information on cost-effective, durable products, reinforcing sustainable consumption habits (Parthiban et al., 2020). By enhancing digital literacy alongside traditional marketplace literacy, these programs can empower BoP consumers to make choices that support both financial stability and long-term well-being, even in the face of income disparity.

3.4 Marketplace Literacy Programs: Building Consumer Competence and Resilience

3.4.1 Marketplace Literacy and BoP Consumers

Marketplace literacy encompasses the essential knowledge, skills, and critical thinking abilities that enable consumers to make informed choices, especially in resource-constrained settings like the BoP. Unlike conventional consumer education, marketplace literacy integrates both practical skills and conceptual understanding, emphasizing the "know-why" behind purchasing decisions alongside the "know-how." This approach helps BoP consumers not only navigate the marketplace effectively but also empowers them to identify and act on their rights as consumers. Given the socio-economic vulnerabilities typical of the BoP segment, marketplace literacy becomes a tool for both consumer protection and empowerment (Viswanathan et al., 2024). In BoP settings, low literacy and a scarcity of resources often limit consumers' abilities to engage in informed decision-making. This makes them more vulnerable to persuasive marketing, which can sometimes drive decisions based on immediate social rewards rather than genuine need. Marketplace literacy aims to close this gap, equipping consumers with practical decision-making tools while fostering a deeper understanding of the market context. Research has shown that marketplace literacy can enhance resilience among BoP consumers, enabling them to make choices that prioritize utility and long-term well-being over status-driven or impulsive purchases.

Studies on ICT's role in the BoP context further underscore that digital access, especially when combined with socio-technical support, can foster the social capital and market access needed to fully realize marketplace literacy's potential. By providing BoP consumers, particularly women, with access to smartphones and community-driven online networks, these programs have been shown to improve structural social capital and economic agency (Parthiban et al., 2022). Through digital access and structured marketplace literacy programs, BoP consumers can develop a more robust understanding of their rights and options, leading to enhanced economic resilience and well-being (Parthiban et al., 2020).

3.4.2 Components of Effective Marketplace Literacy Programs

Marketplace literacy programs are structured to build foundational consumer knowledge while addressing the unique socio-economic and cognitive constraints faced by BoP consumers. At their core, these programs combine practical skills, such as understanding product value and making cost–benefit analyses, with conceptual tools that foster a deeper understanding of the market ecosystem. For BoP consumers, who often lack formal education and have limited exposure to consumer rights and value assessment, these programs provide a transformative learning experience that cultivates both consumer competence and resilience (Jaikumar & Sharma, 2021).

The emphasis on both the "know-how" (technical skills) and "know-why" (underlying concepts) distinguishes marketplace literacy from more conventional consumer education. Effective programs begin with concrete, real-world scenarios and gradually introduce abstract principles, following a bottom-up approach that has proven particularly effective among low-literate, low-income consumers (Viswanathan et al., 2024). For instance, teaching BoP consumers to calculate unit prices or recognize the benefits of bulk buying can empower them to make more cost-effective decisions, while lessons on product durability can shift their purchasing focus from short-term desires to long-term value.

Cultural nuances play a critical role in shaping consumption patterns within BoP communities. For example, in rural India, collectivist values often drive purchasing decisions, whereas in urban BoP settings, individual aspirations may dominate. Marketplace literacy programs must therefore adapt to these cultural differences by tailoring content to reflect local traditions and social dynamics. Regional variations in the implementation of these programs, such as community-led workshops in South Asia versus mobile-based learning platforms in Sub-Saharan Africa, can offer insights into their effectiveness across diverse socio-cultural contexts.

Digital access, such as through smartphones and community ICT initiatives, has also shown promise in reinforcing the practical components of marketplace literacy. Studies indicate that when digital literacy and marketplace skills are taught together, BoP consumers gain even greater confidence in their financial choices and show increased participation in local markets (Parthiban et al., 2020). ICT-enabled learning environments allow consumers to engage with both local and broader marketplaces, thereby building resilience and autonomy in purchasing behaviors that enhance long-term well-being (Parthiban et al., 2022).

3.4.3 Success Stories: Enhancing Resilience During Financial Shocks

One of the most compelling outcomes of marketplace literacy programs is the increased resilience they foster in the face of economic shocks. For example, during the 2016 demonetization in India, marketplace literacy proved instrumental in helping low-income consumer-entrepreneurs adapt to abrupt financial disruptions. The marketplace literacy program designed by Viswanathan et al. (2024) provided essential skills that allowed BoP consumers to manage their limited cash flow, avoid exploitative practices, and find alternative means of transacting, such as bartering. This resilience was particularly significant for women entrepreneurs who, after participating in literacy programs, displayed greater adaptability and resourcefulness in sustaining their micro-enterprises amid cash shortages and systemic challenges.

This example highlights how marketplace literacy can equip BoP consumers with critical coping mechanisms that are essential during periods of economic volatility. By fostering a deeper understanding of marketplace dynamics and financial management, literacy programs transform BoP consumers from passive market participants into active, informed agents capable of making decisions that enhance both their immediate well-being and long-term stability (Mishra et al., 2021). The ability to adapt during financial disruptions not only improves individual resilience but also contributes to economic stability within BoP communities, where financial shocks can otherwise lead to widespread instability.

Digital literacy has been similarly effective in building financial resilience. For instance, digital literacy programs combined with marketplace education have enabled consumers to access digital payment systems, further buffering them from disruptions associated with cash scarcity and enhancing their ability to transact securely (Parthiban et al., 2021). This digital competence has proven especially impactful in BoP markets, where such skills promote both social and financial resilience.

3.4.4 Shifting Conspicuous to Sustainable Consumption Through Literacy

Marketplace literacy programs are instrumental in shifting consumption patterns at the BoP from conspicuous, status-driven purchases to sustainable, need-based choices. By teaching consumers to assess long-term utility over short-term visibility, these programs encourage informed decision-making aligned with economic and environmental sustainability (Jaikumar & Sarin, 2015). For example, consumers are introduced to concepts such as price-per-unit, bulk purchasing, and product durability, enabling them to recognize the financial and practical benefits of durable goods over frequently replaced, status-driven items (Viswanathan et al., 2021).

Digital tools integrated into literacy programs further reinforce these shifts by providing access to real-time price comparisons and product reviews. These resources support informed decisions that prioritize value over visibility, helping BoP consumers reduce waste and allocate resources more effectively (Parthiban et al., 2022). The ability to balance immediate social pressures with long-term benefits is a transformative outcome of marketplace literacy, fostering resilience and sustainable consumption patterns across diverse contexts.

3.4.5 Building Broader Skills Through Marketplace Literacy Programs

Beyond shifting consumption patterns, marketplace literacy programs foster essential skills that empower BoP consumers to navigate complex market environments. Key among these are metacognitive and social intelligence skills, which enable consumers to critically assess their needs, evaluate options, and resist manipulative marketing tactics. Metacognitive skills—such as planning, monitoring, and reflecting—help consumers become more intentional in their decision-making processes, particularly in high-pressure situations (Viswanathan et al., 2021). For instance, participatory learning exercises encourage consumers to identify and reflect on impulsive behaviors, replacing them with thoughtful, value-driven choices.

Social intelligence, which involves understanding and responding to social dynamics, is equally critical in BoP contexts. Consumers often face intense social pressures to conform to conspicuous consumption norms, making it essential to develop strategies to resist these influences. Group discussions and role-playing activities within literacy programs enable participants to practice navigating these dynamics, empowering them to prioritize essential goods and services over status-driven purchases (Parthiban et al., 2021).

Digital literacy complements these broader skills by providing BoP consumers with tools to access reliable information, validate product claims, and engage confidently with online marketplaces. This integration enhances their capacity to make informed, sustainable decisions, aligning personal consumption with broader social and environmental goals (Parthiban et al., 2022).

3.4.6 Measuring the Social and Environmental Impact of Marketplace Literacy

The success of marketplace literacy programs can be gauged not only by individual economic benefits but also by their social and environmental impacts. In terms of social impact, marketplace literacy has been shown to contribute to greater economic stability within BoP communities by reducing the prevalence of status-driven, conspicuous consumption and encouraging responsible spending. This economic resilience has ripple effects, supporting poverty alleviation and enhancing overall community well-being (Viswanathan et al., 2021).

Environmentally, the shift from conspicuous to sustainable consumption reduces waste and promotes the use of durable, long-lasting goods. Studies indicate that when consumers prioritize value-based purchases, they are more likely to select items with longer lifespans and lower replacement rates, leading to reductions in resource usage and environmental impact (Viswanathan et al., 2024). This outcome aligns closely with the Sustainable Development Goals (SDGs), particularly those focused on responsible consumption and production. By equipping BoP consumers with the tools to make sustainable choices, marketplace literacy programs support global environmental objectives while fostering local resilience.

ICT-based interventions also allow for new metrics in evaluating program success. Digital consumption data can be used to track shifts toward sustainable product choices and gauge changes in purchasing behaviors over time (Parthiban et al., 2022). Additionally, digital literacy initiatives enable longitudinal tracking of participants' economic resilience, particularly their capacity to withstand financial shocks and to engage consistently in responsible consumption practices.

Empirical studies demonstrate the measurable outcomes of marketplace literacy programs. For instance, during India's demonetization, participants in literacy programs reported a significant reduction in financial stress and an increase in their ability to navigate alternative transaction methods, such as digital payments and bartering. These outcomes underline the critical role of such programs in fostering economic resilience. Furthermore, longitudinal studies tracking participants over three years indicate sustained improvements in financial management, increased savings, and reduced reliance on high-interest debt. To effectively measure these impacts, program evaluations can incorporate metrics such as changes in spending patterns, increases in durable purchases, and reductions in high-frequency, low-utility purchases. Additionally, longitudinal studies can track the long-term economic outcomes for program participants, including financial stability and resilience in the face of economic challenges (Mishra et al., 2021; Jaikumar & Sharma, 2021). Such comprehensive evaluations are essential for understanding the full potential of marketplace literacy programs to transform BoP markets and contribute to a sustainable future.

3.5 Ethical and Social Implications of Sustainable Marketing to the BoP

3.5.1 Ethical Considerations in Marketing to BoP Consumers

Marketing to BoP consumers requires a high level of ethical consideration due to the socio-economic vulnerabilities that characterize this segment. Many BoP consumers, due to limited literacy, low income, and restricted access to information, are more susceptible to exploitation through persuasive or deceptive marketing tactics. For example, promoting products with exaggerated claims or fostering desires for conspicuous consumption among low-income consumers can lead to financial strain and perpetuate cycles of poverty (Jaikumar & Sarin, 2015). In many cases, such tactics push BoP consumers toward debt as they attempt to keep up with social expectations or satisfy perceived needs for status-driven purchases (Jaikumar & Sharma, 2021). This debt burden can exacerbate their economic challenges, trapping them in a cycle of borrowing and repayment that limits their capacity for sustainable consumption and long-term financial security.

A primary ethical concern in BoP marketing is the potential for exploitation under the guise of aspirational marketing. Campaigns that create an artificial sense of need for luxury or branded items, particularly through aspirational or social comparison messaging, can pressure BoP consumers into spending on products that do not meet their immediate needs or align with their economic capacities. Research indicates that such aspirational marketing often results in high levels of debt among BoP consumers, particularly in rural settings, where social comparison pressures are stronger (Jaikumar & Sharma, 2021). Ethical marketing practices, by contrast, should prioritize products that genuinely enhance the welfare of BoP consumers, focusing on affordability, durability, and utility. For instance, promoting basic health and hygiene products, educational tools, or affordable financial services can improve quality of life and contribute to sustainable development goals.

Marketers must also consider the cognitive and informational constraints of BoP consumers. Low literacy and limited access to comprehensive product information can impair decision-making, leading consumers to make purchases based on incomplete or misleading information (Mishra et al., 2021). Ethical marketing, therefore, includes transparency in product offerings and a commitment to educating BoP consumers on the benefits and limitations of products. Providing clear, accessible information—such as using visuals or simplified explanations—can help BoP consumers make informed choices. In this way, marketing becomes an empowering force rather than an exploitative one, supporting BoP consumers in making decisions that contribute positively to their welfare and mitigate the need to incur debt.

Studies also suggest that incorporating digital literacy and marketplace intelligence into marketing strategies for BoP consumers enhances their resilience against misleading advertising. Digital literacy, for instance, can help BoP consumers navigate online marketplaces more effectively, identify trustworthy sources, and make informed purchasing decisions (Parthiban et al., 2022). This underscores the need for ethical marketing practices that promote informed consumer empowerment through digital education, reducing the likelihood of debt-incurring, impulsive purchases.

3.5.2 Designing Socially Responsible Marketing Strategies for BoP Markets

Developing socially responsible marketing strategies for BoP consumers requires a balanced approach that encourages sustainable consumption without compromising the dignity and autonomy of consumers. One effective strategy is to design marketing campaigns that emphasize the intrinsic value and utility of products, rather than promoting them as symbols of social status. Findings from the study on conspicuous consumption and income inequality show that BoP consumers often prioritize visible goods as a way to signal status, driven by social comparison pressures (Jaikumar et al., 2018). Unfortunately, this focus on visible consumption often drives BoP consumers to take on debt in order to afford status-enhancing products, a practice that can strain their financial resources and reduce their resilience to economic shocks (Jaikumar & Sharma, 2021, Jaikumar et al., 2021). By shifting the messaging to focus on the functional and long-term benefits of products, marketers can help BoP consumers make choices that align more closely with their practical needs and financial well-being.

Marketplace intelligence, as discussed in studies on ethical consumption, provides a framework for creating marketing messages that encourage responsible and need-based consumption (Mishra et al., 2021). For example, instead of promoting short-lived, high-visibility items, campaigns can highlight durable goods or essential services that offer sustained value over time. Marketing strategies that encourage BoP consumers to consider the longevity, value for money, and overall impact of their purchases can facilitate a shift from conspicuous to sustainable consumption, reducing the temptation to incur debt for non-essential purchases.

Furthermore, socially responsible marketing should be sensitive to the unique cultural and social dynamics within BoP communities. Incorporating community-centric messaging that reflects shared values, such as family well-being or collective prosperity, can resonate more effectively and encourage a focus on essential goods. Campaigns that engage local community leaders or trusted voices can also enhance credibility and build consumer trust. For instance, marketing household goods or educational resources by emphasizing their role in family health or community growth aligns with the values of BoP consumers while fostering a sense of shared benefit and sustainable development.

Additionally, socially responsible marketing should avoid creating dependency on products or services that are not locally sustainable. Instead, marketers should seek to support local entrepreneurship and supply chains, promoting products that can be sourced or produced within the community. By supporting local economies, these strategies not only ensure product accessibility but also contribute to sustainable economic growth within BoP markets. Companies can, for instance, provide educational resources that encourage BoP consumers to invest in local goods, reinforcing the value of sustainability through the promotion of community resilience and economic stability (Parthiban et al., 2021).

3.5.3 Sustainable Consumption and the Role of Market Interventions

While marketing strategies play a significant role in shaping consumer behavior, policy interventions are essential to foster sustainable consumption habits among BoP consumers on a broader scale. Market interventions can help mitigate the impact of income inequality, cognitive constraints, and debt by creating supportive conditions for informed, responsible consumption.

One effective intervention is the implementation of policies that promote marketplace literacy programs. As research on consumer empowerment through marketplace literacy has shown, these programs enhance BoP consumers' ability to critically evaluate products and make value-based decisions (Sharma & Jaikumar, 2023). By offering structured learning opportunities that build consumer knowledge and confidence, these programs can empower consumers to prioritize essential goods over conspicuous items, thus fostering sustainable consumption and helping to reduce the likelihood of incurring debt for non-essential purchases. Government and non-governmental organizations (NGOs) can partner with companies to provide educational resources that teach consumers how to evaluate product quality, assess long-term value, and understand the implications of their consumption choices. Such initiatives can be especially impactful in rural or underserved communities, where access to information is limited.

Another important policy measure is the regulation of advertising and labeling standards for products marketed to BoP consumers. Clear and accurate product labeling, especially for essential goods, can help consumers make informed choices without being misled by exaggerated or deceptive claims. For example, labeling that highlights durability or environmental impact can support sustainable consumption by guiding consumers toward products that offer better long-term value. Policy measures can also include restrictions on marketing tactics that exploit cognitive biases, such as scarcity messaging or emotional appeals, which can disproportionately impact consumers with limited access to alternative sources of information. By ensuring transparency and clarity in product information, these policies can reduce the incidence of debt-fuelled purchases for non-essential goods (Jaikumar & Sharma, 2021).

Financial inclusion policies can also support sustainable consumption. For example, policies that expand access to microfinance or savings programs allow BoP consumers to build financial security and make more considered consumption choices. As demonstrated in studies on marketplace literacy, consumers with greater financial literacy and access to financial services are more resilient to economic shocks and less susceptible to exploitative marketing. By supporting financial literacy alongside marketplace literacy, policymakers can help BoP consumers develop the capacity to allocate resources toward products that improve quality of life and foster economic resilience, reducing the reliance on debt as a means to fulfil consumption needs.

Finally, collaborative partnerships between governments, NGOs, and private companies can create a comprehensive support system for sustainable consumption at the BoP. For instance, companies can work with NGOs to develop products that address local needs sustainably, while governments can facilitate the distribution of these products through subsidies or reduced taxes on essential goods. Such collaborations can also encourage the development of environmentally sustainable products, aligning both corporate and governmental objectives with community welfare. By creating a supportive ecosystem for sustainable consumption, these interventions address the systemic barriers that BoP consumers face, empowering them to make choices that support their long-term well-being and contribute to sustainable development goals.

In sum, ethical and socially responsible marketing, coupled with targeted policy interventions, can foster a sustainable consumption mindset within BoP markets. By prioritizing consumer welfare, promoting informed choice, and implementing supportive regulations, marketers and policymakers can work together to create a marketplace environment that empowers BoP consumers, supports economic resilience, and aligns with broader sustainability goals.

3.6 Practical Applications: Marketing Strategies and Policy Recommendations

3.6.1 Marketing Strategies for Encouraging Sustainable Consumption

3.6.1.1 Promote Value-Based, Durable Products Over Status-Driven Goods

Marketing strategies should prioritize the functional and long-term benefits of products over status-driven attributes. Emphasizing value, utility, and durability aligns with the economic realities of BoP consumers, helping them make informed choices that serve their practical needs. Campaigns can leverage product features that BoP consumers find useful, such as affordability, durability, and versatility, to encourage sustainable consumption (Jaikumar et al., 2018; Viswanathan et al., 2024). For example, marketers could highlight cost savings achieved through bulk purchases or showcase products that reduce frequent replacement needs. Emphasizing the value of essentials—such as health, education, and hygiene products—can address the tendency toward conspicuous consumption, which often leads to debt (Jaikumar & Sharma, 2021). Marketing that centres around these sustainable attributes can foster a shift in BoP spending from short-term, status-driven purchases to long-term, need-based choices.

3.6.1.2 Integrate Digital Literacy into Marketplace Literacy Programs

To further support sustainable choices, digital literacy should be integrated into marketplace literacy programs, particularly as digital access grows within BoP markets. With smartphones and other ICT tools becoming more prevalent, marketers can provide digital resources, such as cost-comparison tools and information on product durability, directly accessible through apps or online platforms. Digital literacy enables consumers to navigate online marketplaces effectively, identify trustworthy information sources, and avoid misleading or exploitative offers (Parthiban et al., 2022). Marketers can work with NGOs and local organizations to create digital tutorials that guide BoP consumers through essential online tasks, like reading product reviews, comparing prices, and recognizing signs of misleading advertising. Digital literacy not only empowers BoP consumers to make sustainable choices but also strengthens their economic resilience by reducing the likelihood of incurring debt for non-essential purchases.

3.6.1.3 Leverage Community-Centric and Culturally Relevant Messaging

Marketing strategies should reflect the values, traditions, and community dynamics of BoP consumers. Culturally relevant messaging that emphasizes collective well-being—such as the importance of family health, community resilience, and education—can resonate more deeply with BoP consumers, reinforcing the value of need-based products over conspicuous consumption. Campaigns can benefit from enlisting local community leaders or respected voices to communicate these messages, enhancing credibility and building consumer trust (Viswanathan et al., 2021). Marketers can also focus on promoting locally produced goods, which can create a sense of community pride and support local economies. This approach aligns with the goals of sustainable consumption by encouraging BoP consumers to invest in products that foster community growth, while reducing reliance on external, non-sustainable products (Parthiban et al., 2021).

3.6.2 Policy Recommendations for Supporting Marketplace Literacy and Sustainable Consumption

3.6.2.1 Implement and Expand Marketplace Literacy Programs with Digital Integration

Policy interventions should prioritize the implementation and expansion of marketplace literacy programs with integrated digital literacy training. Government and non-governmental organizations (NGOs) can collaborate with private companies to develop comprehensive programs that teach both practical and digital skills, equipping BoP consumers with the tools to navigate marketplaces confidently and sustainably. Evidence suggests that programs combining financial and digital literacy result in greater economic resilience and more sustainable purchasing behaviors among BoP consumers (Sharma & Jaikumar, 2023). Governments can facilitate the development and accessibility of mobile apps and online platforms specifically designed for BoP consumers, which offer budgeting tools, product comparisons, and educational content on sustainable consumption. By providing the technological infrastructure and necessary training, these initiatives can empower BoP consumers to make informed decisions that align with their financial realities and long-term goals.

3.6.2.2 Regulate Advertising to Protect Vulnerable Consumers

Regulation of advertising, especially digital advertising, is crucial to protect BoP consumers from exploitative marketing practices. Policies should enforce transparency in advertising, mandating clear information on product durability, quality, and environmental impact. Misleading advertising that emphasizes luxury or status over utility can drive BoP consumers into debt, as they may prioritize conspicuous purchases over essentials (Jaikumar & Sharma, 2021). Additionally, restrictions on high-pressure marketing tactics, such as scarcity messages or emotional appeals, can help shield BoP consumers from impulsive buying. Regulatory bodies should consider establishing guidelines that require companies to emphasize product value, quality, and relevance to essential needs, which can aid BoP consumers in making responsible and sustainable choices.

3.6.2.3 Develop Financial Inclusion Programs and Accessible Savings Solutions

Policymakers can support sustainable consumption through financial inclusion programs, such as microfinance, low-interest loans, and accessible savings plans. Financial inclusion not only provides BoP consumers with the capital needed to make larger, durable purchases but also helps them avoid high-interest debt cycles associated with informal loans. Research has shown that BoP consumers with access to savings and financial literacy programs are more resilient to economic shocks and make more thoughtful purchasing decisions. Savings programs can be particularly effective when paired with marketplace literacy, as consumers learn to allocate their resources toward products that improve quality of life and financial stability. By fostering savings habits, BoP consumers are better positioned to make purchases that serve their long-term needs, reduce environmental impact, and support sustainable growth within their communities.

3.6.2.4 Foster Partnerships Between Government, NGOs, and Private Sector

Collaborative partnerships are essential to create a supportive ecosystem that promotes sustainable consumption. Governments, NGOs, and private companies can work together to develop and distribute sustainable products designed for BoP consumers, as well as to expand educational initiatives. These partnerships can offer subsidies or reduced taxes on essential goods, making sustainable products more affordable to BoP consumers and encouraging widespread adoption. For example, NGOs can assist in developing community-based educational programs, while private companies provide affordable and locally sourced goods. Governments can further incentivize companies to adopt sustainable production practices, benefiting both consumers and the environment. Such partnerships foster a cohesive system that addresses the multiple barriers BoP consumers face, allowing them to make choices that are aligned with long-term sustainability and community welfare (Parthiban et al., 2021).

Despite its transformative potential, ICT adoption among BoP consumers faces significant barriers. Affordability of devices, poor digital infrastructure in rural areas, and gaps in digital literacy remain key challenges. To address these barriers, targeted policies such as subsidies for low-cost devices, investments in rural internet infrastructure, and community-based digital literacy programs are essential. For instance, public–private partnerships can facilitate the development of affordable internet services in underserved areas, bridging the digital divide and enabling greater access to ICT-based marketplace literacy interventions.

3.7 Conclusion: Toward a Sustainable Future for BoP Markets

The Base of the Pyramid (BoP) market segment represents a substantial population with untapped potential for economic growth, sustainability, and social development. This chapter has explored the unique challenges faced by BoP consumers—including limited literacy, cognitive constraints, social pressures, and vulnerability to exploitative marketing practices—that hinder informed and sustainable decision-making. Addressing these challenges requires comprehensive, tailored interventions that promote marketplace literacy, sustainable consumption, and resilience. Marketplace literacy, as a transformative approach, empowers BoP consumers by equipping them with critical knowledge, practical skills, and digital literacy, enabling them to navigate complex market environments with confidence and purpose.

Marketplace literacy programs, as examined in this chapter, play a crucial role in building both consumer competence and resilience. Through education on practical decision-making, cost–benefit analyses, and sustainable choices, these programs enable BoP consumers to shift from conspicuous to need-based consumption. Additionally, digital literacy—now an integral part of marketplace education—provides BoP consumers with access to a broader range of resources and information, enhancing their capacity for informed decision-making and reducing susceptibility to debt-driven, impulsive purchases.

Ethical marketing and policy interventions are equally vital in shaping a sustainable consumption environment for BoP consumers. Responsible marketing that prioritizes value, affordability, and durability, combined with community-centric messaging, can help align BoP consumption patterns with long-term well-being. Additionally, policies that support marketplace literacy, regulate advertising practices, and promote financial inclusion create supportive frameworks for BoP consumers, enabling them to make sustainable and economically sound choices. Collaborative partnerships among governments, NGOs, and private companies further amplify these efforts, ensuring the availability of affordable, sustainable products tailored to the needs of BoP communities.

Looking ahead, creating a sustainable future for BoP markets requires sustained commitment from all stakeholders. Expanding marketplace literacy programs, strengthening digital infrastructure, promoting culturally relevant marketing practices, and fostering collaborative partnerships are essential steps. Through these combined efforts, BoP consumers can achieve economic resilience, contribute to sustainable development, and forge pathways out of poverty.

By embracing a holistic approach that respects the unique needs and aspirations of BoP consumers, we can foster an inclusive and sustainable marketplace that uplifts individuals, strengthens communities, and advances broader environmental and social goals. This vision of sustainable growth is not only achievable but essential for addressing the world's most pressing challenges and creating a future that benefits all.

References

Jaikumar, S., & Sarin, A. (2015). Conspicuous consumption and income inequality in an emerging economy: Evidence from India. *Marketing Letters, 26*, 279–292.

Jaikumar, S., & Sharma, Y. (2021). Consuming beyond means: Debt trap of conspicuous consumption in an emerging economy. *Journal of Marketing Theory and Practice, 29*(2), 233–249.

Jaikumar, S., Dutta, S., & Sood, N. (2021). Impact of lifestyle diseases on income and household consumption: Evidence from an emerging economy. *Health Marketing Quarterly, 38*(1), 35–49.

Jaikumar, S., Sharma, Y., & Silal, P. (2020). Impact of education on health-beliefs in an emerging economy: Understanding urban vs. Rural differences and the moderating effect of social capital. *Health Marketing Quarterly, 37*(4), 316–332.

Jaikumar, S., Singh, R., & Sarin, A. (2018). 'I show off, so I am well off': Subjective economic well-being and conspicuous consumption in an emerging economy. *Journal of Business Research, 86*, 386–393.

Mishra, R., Singh, R., & Jaikumar, S. (2021). Executive functions of BoP consumers: Research propositions, conceptual framework and implications for marketing strategies for BoP markets. *Journal of Global Marketing, 34*(4), 249–269.

Parthiban, R., Jaikumar, S., Basak, J., & Bandyopadhyay, S. (2022). Digital access through smartphones and well-being of BoP women: Insights from a field study in India. *Information Technology & People, 35*(1), 1–26.

Parthiban, R., Qureshi, I., Bandyopadhyay, S., & Jaikumar, S. (2021). Digitally mediated value creation for non-commodity base of the pyramid producers. *International Journal of Information Management, 56*, 102256.

Parthiban, R., Qureshi, I., Bandyopadhyay, S., Bhatt, B., & Jaikumar, S. (2020). Leveraging ICT to overcome complementary institutional voids: Insights from institutional work by a social enterprise to help marginalized. *Information Systems Frontiers, 22*, 633–653.

Sharma, Y., & Jaikumar, S. (2023). Educational approaches to enable marketplace metacognition and social intelligence in subsistence consumers. *Journal of Consumer Marketing, 40*(1), 44–55.

Viswanathan, M., Jaikumar, S., Sreekumar, A., & Dutta, S. (2021). Marketplace literacy education and coping behaviors among subsistence consumer–entrepreneurs during demonetization in India. *Journal of Consumer Affairs, 55*(1), 179–202.

Viswanathan, M., Jaikumar, S., Sreekumar, A., Dutta, S., & Duhachek, A. (2024). Addressing difficulties with abstract thinking for low-literate, low-income consumers through marketplace literacy: A bottom-up approach to consumer and marketing education. *Journal of Consumer Affairs, 58*(3), 794–813.

Saravana Jaikumar is a faculty member of the Marketing Group at IIM Calcutta. His research work has been published in reputed journals including Journal of Marketing, Journal of Advertising Research, Journal of Business Research, Journal of Consumer Affairs, Marketing Letters and others. His teaching is mainly in the areas of digital marketing, marketing technology and analytics. He completed his doctorate from IIM Ahmedabad, MBA from Cardiff Business School, UK and BE from PSG Tech, Coimbatore.

Chapter 4
Bridging Economic, Social, and Environmental Sustainability in BoP Markets: An Integrated Approach

Prama Vishnoi, Anupama Vohra, and Neha Bhardwaj

Abstract Despite efforts to promote sustainable consumption, unsustainable practices persist globally, particularly in Bottom-of-the-Pyramid (BoP) markets where environmental degradation and social inequalities disproportionately affect the world's poorest populations. This chapter addresses this challenge by adapting Lim's Sustainability Pyramid approach to BoP contexts. The chapter proposes that sustainability initiatives in BoP markets should prioritize economic sustainability, followed by social and environmental dimensions. This hierarchical approach aligns with BoP realities, where economic concerns often take precedence. The chapter explores how academics, policymakers, and firms can collaboratively influence consumer decision-making toward sustainable consumption in BoP markets using this adapted framework. The implications are significant: for theory, it offers a new lens for examining sustainability in BoP markets; for public policy, it suggests a tiered approach to sustainability initiatives; for business practice, it provides guidance on developing economically viable, socially responsible, and environmentally conscious solutions for BoP markets. This model has the potential to drive meaningful, long-term sustainable development in BoP communities, addressing both immediate economic needs and broader sustainability challenges while recognizing the diverse contexts within BoP markets globally.

Keywords Sustainability · Bottom of the Pyramid · Sustainable Consumption · Consumer Behavior · Economic Sustainability · Social Sustainability · Environmental Sustainability · BOP

P. Vishnoi
Keshav Mahavidyalaya, University of Delhi, Delhi, India
e-mail: pramavishnoi.17@fms.edu

A. Vohra · N. Bhardwaj (✉)
Faculty of Management Studies, University of Delhi, Delhi, India
e-mail: nehabhardwaj@fms.edu

A. Vohra
e-mail: anupamavohra@fms.edu

R. Singh and T. A. Wani (eds.), *Marketing to the Base of the Pyramid*,
https://doi.org/10.1007/978-981-96-7105-2_4

4.1 Sustainability and Its Importance

In recent years, the concept of sustainability has gained significant importance worldwide. It has emerged as a prominent term used by consumers, organizations, and nations to evaluate the impact of human activities on the ecosystem (Lim, 2022). It has often been referred to as the most significant translational problem of today's times. Moore et al. (2017) noted for most literature on sustainability to be theoretical while focussing less on the delivery and implementation of the same. The scholars also note a variety of definitions and terminology to be given to sustainability in the literature. Based on their detailed theoretical assessment of literature, Moore et al. (2017) identified five key constructs to be dimensionalizing sustainability across literature: (1) Time—a defined period of time, (2) Continued Delivery—the program implementation/strategies continue to be delivered, (3) Behavior Change—Individual behavior change is maintained, (4) Evolution—the program / individual behavior change may adapt/evolve, and (5) continued benefits—continuing to produce benefits for individuals/systems. While these concepts may be distinct, they tend to be interrelated.

A merely growth-oriented paradigm to marketing has today slipped into a blind alley. Marketing practices today emanate from an "Emergent Marketing Paradigm" which focuses on sustainable and sufficiency marketing exposing itself to an alternative marketing system (Kelleci & Yildiz, 2021). Academic literature is bursting with sustainability definitions both within and outside the context of marketing. The concept of sustainability is a multi-dimensional concept, and each of its dimensions has led to a different discourse everytime. The individual discourse adopted for each of its dimensions has limited the actual implementation of sustainability to a mere rhetoric. The ubiquity of the concept has also led to the association of sustainability with social responsibility, environmental management, or business sustainability, which are often treated separately. However, there is a need for an integrated approach to be adopted at the systems level.

The multidimensionality of sustainability was identified through the Rio + 20. The Rio + 20 outcome document referred to three dimensions of sustainability: economic, social, and environmental. Further, it also identified good governance as an overarching basis for sustainable development. In a similar light, the Global Reporting Initiative (GRI), has highlighted for the various dimensions of sustainability to be disclosed within reporting activities.

There are hundreds of definitions of sustainability in the literature. While some consider the multiplicity of definitions as a source of confusion, a few others take it as an advantage. In this regard, White (2013) indicated for different interpretations of sustainability to be derived from these definitions, while Costanza and Patten (1995) called for this interpretation to focus on what is to be sustained. Hahn (2009) identified for the fundamental elements of sustainable development to be based on intergenerational and intragenerational justice. While intergenerational justice was described to be focussing on the ecological perspective, the focus of intragenerational focuses on social and economic adjustments. In this sense, the scholars described sustainability as an ethical decision rather than a law of nature.

Far from diminishing, sustainability concerns have grown more complex over time. This evolution is evident in the expansion from eight United Nations Millennium Development Goals (UN MDGs) for 2000–2015 to 17 United Nations Sustainable Development Goals (UN SDGs) for 2016–2030 (Lim, 2022). Recent assessments of SDG progress in the Asia–Pacific region suggest that many nations are unlikely to meet most of the 2030 SDG targets. The COVID-19 pandemic has further complicated these efforts, as highlighted in the Sustainable Development Goals Report 2021. Notably, global extreme poverty increased for the first time in over two decades, affecting an additional 119 to 124 million people (Lim, 2022; United Nations Department of Economic & Social Affairs, 2021).

In light of these challenges and the growing complexity of sustainability issues, there has been a significant shift in the approach to addressing global poverty. The private sector is increasingly being called upon to participate in poverty alleviation efforts (Achrol & Kotler, 2016; Rahman et al., 2015; Scheyvens et al., 2016). This approach received formal endorsement from the United Nations in 2015 with the adoption of the 17 Sustainable Development Goals, which encourage proactive private sector engagement in eradicating extreme poverty by 2030 (United Nations, 2016). This recognition underscores the critical role businesses can play in addressing one of the world's most pressing challenges (Voola et al., 2024; Voola & Voola, 2019).

This emphasis on sustainability, shared by governments, non-governmental organizations (NGOs), and businesses, is mirrored in academic discourse. Scholars argue that studying sustainability is essential as it forms the context shaping key phenomena of our century (e.g., McDonagh & Prothero, 2014).

4.2 Conceptual Background

Approximately half of the global population lives in poverty, facing resource scarcity due to limited incomes. While their individual purchasing power pales in comparison to that of the wealthy, their collective buying potential forms a substantial market. These impoverished individuals, referred to as the "bottom of the pyramid" (BoP) or "base of the pyramid," represent a crucial market for businesses worldwide (Prahalad, 2005). With an estimated 4 billion people earning annual incomes below US$1,500 (based on purchasing power parity), this segment commands a combined annual purchasing power exceeding US$5 trillion (Hammond et al., 2007), making it an attractive market for businesses to consider.

The conceptualization of sustainability in BoP markets necessitates recognition of two distinct consumption patterns: market-dependent and nature-dependent (Venugopal & Chakrabrati, 2022; Venugopal et al., 2024). In market-dependent BoP contexts, consumption patterns are predominantly shaped by market mechanisms and economic transactions. Conversely, in nature-dependent BoP communities, such as coastal fishing communities or forest-dwelling populations, livelihoods are inextricably linked to natural resources, making the dimensions of sustainability fundamentally interconnected. For these communities, environmental sustainability is not merely a tertiary concern but is deeply intertwined with economic survival and social structures.

The significance of studying BoP markets is amplified by demographic trends and sustainability concerns. As the majority of the world's population growth occurs in this segment, the BoP consumer base is projected to reach 6 billion in the next four decades (Prahalad & Hart, 2002). However, this growth presents both opportunities and challenges. While it represents a latent market with exceptional sustainable growth potential, there are concerns that if these new consumers adopt the consumption practices of developed economies, it could lead to severe resource depletion and environmental degradation (Viswanathan & Rosa, 2010). Higher consumption levels necessitate increased production, requiring more significant inputs and resulting in larger quantities of waste and by-products (Vohra & Vishnoi, 2022).

The sustainability challenges in BoP markets are particularly acute due to their unique socio-economic features. Rieckmann (2013) highlighted the Eurocentric focus of sustainability discourse, noting the absence of perspectives from Africa, Latin America, and Asia in sustainability discussions (Demssie et al., 2019). These unique features necessitate specific sustainability competencies and BoP-focused frameworks that prioritize dimensions of sustainability based on socioeconomic context (Demssie et al., 2019). Notably, Hahn (2009) positions sustainability at the BoP as a claim for intragenerational justice, encompassing both social and economic justice.

Addressing these challenges requires a multi-stakeholder approach. The responsibility extends beyond government policies and initiatives to include business organizations, which must play a crucial role in creating value for BoP consumers while preventing irreversible environmental damage. Conceptual and review studies are vital in reshaping research practices and enhancing our ability to tackle both theoretical and practical problems in sustainability marketing (Lim, 2022). By targeting academics, firms, and governments, this chapter seeks to improve sustainability marketing efforts, driving significant behavioral changes and reducing the environmental impact of consumption in BoP markets (Zhao et al., 2016).

4.3 Suitability of the Sustainability Pyramid for BOP

Sustainability has become a popular concept for evaluating human impact on the environment, but many efforts in this area have been lacklustre. Sustainable products often fail to meet consumer expectations, being either overpriced, inconvenient, or ineffective. Although some consumers show willingness to adopt more sustainable practices, they remain a minority. Most people readily change their habits for personal benefit, such as adopting new communication technologies or fashion trends, but are less inclined to make changes for the greater good. To drive wider adoption of sustainable practices, it's crucial to demonstrate how these choices can enhance individual well-being, rather than solely focusing on societal or future benefits (Lim, 2022).

Sustainability has been most often described in terms of its three dimensions—economic, social, and environmental. However, in the context of BOP, it has implications on how the three elements need to be balanced. Several studies have had differing suggestions on the same. While Ngosso (2013) indicated that both economic and environmental focus may lead to poverty reduction, Tomsic et al. (2015) call for equal priority to be given to all three elements.

Giovannoni and Fabietti (2013), highlighted the tensions that exist between the various sustainability dimensions and the challenges with developing an integrated approach to sustainability. Gray (2010) suggested a sustainability continuum between strong and weak sustainability, as a means to address the tensions between the three dimensions. These approaches, thus, indicate for all dimensions to be managed simultaneously through an integrated approach. To this end, Giovannoni and Fabietti (2013) suggested a three-level framework to implementing integrated sustainability—at the basic level lies Management, measurement and reporting systems, level 2 is strategy and business models and integrated governance from the top level. The framework, therefore, presents an organization focussed approach to sustainability—highlighting them as the key stakeholders toward achieving sustainability in the long run. (this takes an organization focussed view, ours is customer focussed; it calls for all dimensions to be managed simultaneously, ours presents a hierarchy).

Kelleci and Yildiz (2021) based on Kotler's Taxonomy of marketing identified a sustainability framework based on levels into which sustainability may be divided based on factors such as scope of marketing, marketing orientation and the platforms adopted (they are more focussed on marketing definition of sustainability and not its implementation as our framework). Hahn (2009) observed that most of the BOP models were narrow in their approach as they failed to discuss how BOP development could happen without creating any disastrous effects on the environment. The scholars called for positive correlations between intergenerational (environmental) and intragenerational (social and economic) justice, in the context of BOP. With the BOP being more dependent on the ecological environment and also carrying the disproportionate environmental burdens—measures to improve the environment may lie at the intersection of social and economic justice. However, what the scholar did postulate was for relationships that exist among the three levels of justice as identified in their study.

Lim's (2022) sustainability pyramid (Fig. 4.1) is a hierarchical model integrating economic, environmental, and societal dimensions of sustainability. It adopts a consumer-centric approach, placing consumers at the core of sustainability strategies. The model suggests addressing lower-level sustainability dimensions before progressing to higher ones, aiming to guide society toward greater overall sustainability. The economic dimension forms the foundation of the sustainability pyramid, as unsustainable consumption by consumers is the root cause of many related problems. Most consumers, especially those living in poverty (over 20% of the world's population), prioritize economic value in their purchasing decisions. The social and

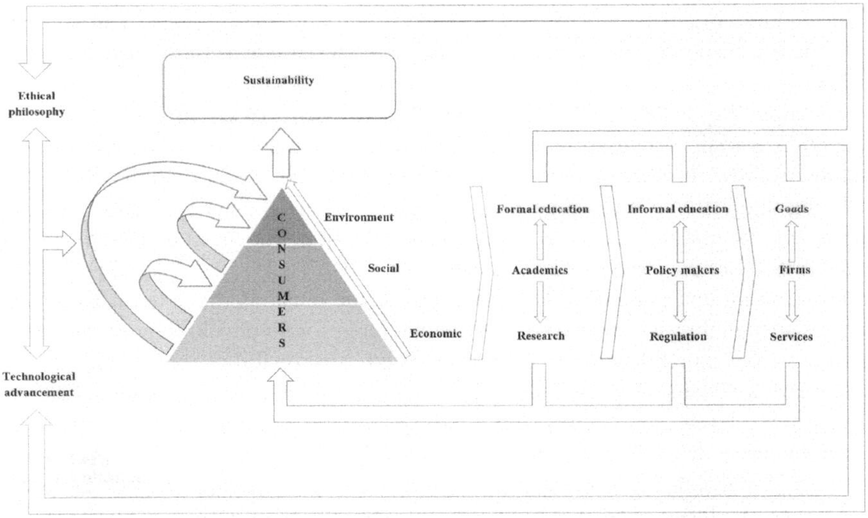

Fig. 4.1 The sustainability pyramid (*Source* Lim, 2022)

environmental dimensions occupy the second and third levels of the sustainability pyramid, respectively. This suggests that consumers must first be convinced of the economic value of sustainable alternatives before introducing social and environmental aspects. Social sustainability aims for equitable, diverse, and democratic communities with a good quality of life, while environmental sustainability seeks to meet human needs without compromising the planet's future capacity. Achieving these dimensions requires intervention from academics (knowledge discovery and dissemination), policymakers (macro-level responsibilities), and firms (micro-level responsibilities). Eventually, the model suggests that ethical philosophy informs and technological progress shapes the pursuit of sustainability across economic, social, and environmental domains.

Interestingly, as Lim's model/pyramid approach emphasizes the importance of addressing economic concerns first, it aligns with the primary needs of BOP consumers. The sustainability pyramid approach is particularly relevant to the BOP and can be adapted to better suit BOP realities by recognizing and leveraging the sustainability awareness and practices within these communities. The pyramid emphasizes addressing economic sustainability first, which aligns with the needs of BoP consumers, who prioritize economic survival over environmental or social concerns. By presenting sustainable products that provide clear economic benefits, such as cost savings or increased income opportunities, businesses can engage this segment effectively. Once economic needs are met, BoP consumers may gradually adopt socially and environmentally sustainable behaviors, creating a foundation for broader sustainable development initiatives. BOP consumers often demonstrate a strong commitment to sustainability through their actions. Despite limited resources, they engage in pro-environmental activities, recognizing the direct impact of environmental quality on their lives. Their willingness to participate in eco-friendly practices stems from both a desire to protect themselves from pollution and a genuine care for the environment. Surprisingly, BOP consumers exert a significant positive influence on sustainability efforts, challenging the notion that sustainability consciousness is limited to more affluent demographics. This phenomenon highlights the potential for inclusive, sustainable development strategies that leverage the sustainability awareness and motivations of BOP communities (Zhao et al., 2016).

4.4 Sustainability Pyramid for the BOP

The sustainability Pyramid, as proposed by Lim (2022), presents a compelling framework for promoting sustainable practices. The model's hierarchical approach prioritizes economic sustainability, followed by social and environmental dimensions. This conceptual framework offers a strategic approach for promoting sustainable development; particularly relevant for application in the context of Bottom of the Pyramid.

The adaptation of the framework to BOP markets requires a nuanced understanding of the unique challenges and opportunities presented by these markets. This chapter explores how the Sustainability Pyramid's three dimensions—economic, social, and environmental can be tailored to the BOP context, while also considering the interventions of three key stakeholders: academics, policymakers, and firms.

4.4.1 Economic

The economic discourse on sustainability often brings forward the paradox between corporations and sustainability (Gray, 2010). While organizations possess the capabilities to produce large scale innovations, and a means to implement sustainability; they also lie at the heart of concerns involving social inequalities and resource deterioration.

At the foundation of the Sustainability Pyramid lies the economic dimension. It prioritizes prosperity, which is highly significant in BOP markets due to severe resource constraints. As such it aligns with the immediate needs of the BOP. Adapting this dimension for BOP consumers necessitates a focus on affordability, immediate cost savings, and long-term economic benefits. By emphasizing on economic benefits first, firms can create products that are affordable and improve living standards. This focus is necessary for engaging BOP consumers, who often prioritize cost and utility due to their limited financial resources. Sustainable economic initiatives for BOP might include microfinance, affordable healthcare facilities and low-cost, high-quality goods tailored specifically to their needs. Here, the interplay between academics, policymakers, and firms becomes critical.

Academics can drive research into cost-effective sustainable solutions specifically designed for BOP markets. Through research, they can identify the most effective economic interventions providing valuable insights that inform both policy and business strategies. They can then disseminate this acquired knowledge by educating both firms and communities on sustainable practices. Firms can leverage these academic insights to innovate business models-such as microfinance initiatives or pay-as-you-go schemes. They can take on micro-level responsibility by developing and delivering sustainable products and making them more accessible to BOP consumers. Policymakers, assuming macro-level responsibilities, can create an enabling environment by formulating policies that support economic activities at the BOP. Armed with academic research, policymakers can create a supportive environment through targeted incentives (such as subsidies, tax incentives) and regulatory frameworks that encourage the development of affordable sustainable products. This collaborative approach can result in sustainable offerings that clearly demonstrate tangible cost savings, a crucial factor for product adoption among BOP consumers. While this economic focus aligns with BOP consumers' primary concerns, it's important to note that it also lays down the groundwork for introducing social and environmental considerations.

Bringing the economic value of sustainability to the forefront for bottom-of-the-pyramid (BoP) consumers involves creating products and services that align with their financial constraints, while simultaneously addressing environmental and social needs. For instance, solar energy solutions, M-KOPA in East Africa offers solar home systems to low-income households on a pay-as-you-go basis. Instead of paying high upfront costs, customers pay small daily installments via mobile money, eventually owning the solar system. By switching to solar energy, BoP consumers save money on kerosene, batteries, and candles, which are expensive and harmful to health. Solar power also provides clean, renewable energy, reducing the environmental impact. The affordability and financing model makes it possible for BoP consumers to invest in clean energy, improving their standard of living while reducing long-term energy costs. The water purification system, PureIT by Unilever India offers low-cost, non-electric water purifiers designed to provide safe drinking water to low-income households in India. Access to clean water helps prevent illness, reduce the use of non-renewable energy (e.g., firewood or gas for boiling), and improve overall household health, making it both a sustainable and cost-effective solution for BoP consumers. The water purifiers eliminate the need for boiling water, saving money on fuel and reducing health costs associated with waterborne diseases. The purifier itself is affordable and has low operating costs.

To sum up, products and services must be affordable and accessible to BoP consumers. Payment models such as microcredit or pay-as-you-go can help reduce the upfront financial burden. Sustainable products should demonstrate clear economic value by reducing ongoing costs for energy, health, or resources, such as fuel-efficient stoves or solar panels. By creating solutions that align with the financial needs of BoP consumers, companies can make sustainability not just an environmental or ethical priority but also an economically viable one for the world's poorest communities.

4.4.2 Social

The Millennium Development Goals established in 2000 after the Kyoto Conference on Climate Change, 1997 laid a key milestone for addressing social concerns. This was followed by the World Summit on Sustainable Development in Johannesburg, 2002, which shifted the focus of sustainability from merely environmental concerns to social and economic development as well (Giovannoni & Fabietti, 2013).

Once the economic benefits are secured and the financial stability of BOP communities is enhanced, social dimension, occupying the middle tier of the Sustainability Pyramid becomes more relevant and can be more effectively addressed. The social aspect puts emphasis on the significance of people, focusing on improving health, education, and overall well-being. In the BOP context, this dimension can highlight how sustainable choices can enhance health, education, and overall quality of life, fostering community resilience. Social initiatives might include: affordable water purification systems improving community health, solar-powered lamps enhancing educational opportunities, and skills training programs for maintaining sustainable technologies. These initiatives can not only address immediate social needs but also contribute to long-term community development, empowerment and social mobility, demonstrating the interconnected nature of social sustainability with economic and environmental benefits in BOP markets. Academics can contribute to this by identifying intricate social structures, cultural norms, and aspirations of BOP communities, providing a knowledge base that informs effective social sustainability initiatives. Firms can then play a crucial role in bringing these ideas to life by leveraging their resources and expertise. They can develop products and services that not only meet economic needs but also foster skill development, create jobs, and promote social mobility within BOP communities. Policymakers can support these efforts by crafting community-based programs that resonate with local values and social networks; informing and educating the BOP consumers on sustainability issues. This collaborative effort ensures that social sustainability initiatives are culturally appropriate, economically viable, and supported by policy. However, it's pertinent that all stakeholders avoid imposing external notions of social sustainability, instead working closely with BOP communities to co-create relevant definitions and metrics of social progress. This will ensure that social advancements are sustainable and impactful. Sustainability efforts should offer not just environmental benefits but also direct improvements in quality of life, health, education, and social equity. Programs that incorporate BoP communities into sustainable supply chains or service models can generate employment and enhance local economic resilience.

For instance, M-KOPA provides access to clean energy, reduces energy poverty, improves health by reducing indoor air pollution, and increases productivity and educational outcomes by extending hours of light. This also creates local jobs in installation and maintenance, leading to economic and social development. PureIT provides access to clean and safe drinking water (In many BoP communities, access to clean and safe drinking water is limited, leading to widespread waterborne diseases such as diarrhea, cholera, and typhoid. By reducing the prevalence of waterborne diseases, Pureit helps improve public health, which directly impacts the quality of life for BoP communities. Healthier individuals are more likely to participate in education, work, and other community activities, promoting overall societal well-being.); affordability (Many traditional water purification systems are too expensive for BoP consumers to afford. Providing an affordable solution to clean water addresses the issue of economic inequality, enabling BoP consumers to access basic human need without placing a financial burden on them. This inclusivity promotes equity in health outcomes and improves long-term community resilience.)

At the social sustainability levels, initiatives should focus on empowering BoP communities by providing them with the tools and resources to improve their own social and economic circumstances. For instance, many sustainability programs address health as a primary social value, improving public health through better sanitation, nutrition, and healthcare access, leading to more sustainable communities. Promoting access to education and digital inclusion ensures long-term social mobility, particularly for children and women, which enhances the social fabric of BoP communities. Sustainable initiatives often tackle multiple social issues—such as poverty, education, health, and employment—simultaneously, creating a more profound and lasting social impact. By highlighting the social value of sustainability, companies and organizations can address critical societal needs while fostering long-term development for BoP consumers, ultimately improving their quality of life.

4.5 Environmental

Sustainability has more often than not been compartmentalized as an environmental issue. The 19070 s saw the term 'sustainability' being widely used in relation to environmental problems (Giovannoni & Fabietti, 2013). The Stockholm Conference of 1972 (UN, 1972) led to the creation of the UN Environmental Programme and a number of environmental protection agencies. The 1970s to 1990s witnessed sustainability primarily in relation to environmental factors.

The environmental dimension becomes increasingly relevant as the economic and social foundations are solidified. It is at the apex of the Sustainability Pyramid. It presents unique opportunities and challenges in the BOP context. While immediate economic concerns often overshadow environmental issues for those living at the BOP, there is often a direct link between environmental degradation and BOP livelihoods. For the BOP, environmental sustainability involves implementing practices and technologies that reduce environmental impact. This could include initiatives like promoting renewable energy, sustainable agriculture, and waste management systems. At the BOP, environmental sustainability can also involve education and awareness programs to foster environmental stewardship within communities. In addressing this dimension, the roles of academics, firms, and policymakers are again interconnected. Academics can provide critical research on environmental impacts and sustainable technologies, ensuring that interventions are scientifically sound and effective. They can conduct research that illuminates the connection between environmental preservation and economic stability in BOP contexts, providing evidence-based insights that drive policy and business decisions. Firms can apply this knowledge in their operations and product offerings, creating environmentally friendly solutions that are accessible to BOP communities. They can innovate to create products and services that address environmental concerns while meeting the economic and social needs of BOP consumers. For example deploying affordable, eco-friendly agricultural inputs that improve soil fertility and crop yields. Policymakers can use academic research to create regulatory frameworks that mandate or incentivize sustainable practices, ensuring that environmental considerations are integrated into broader development strategies. Specifically, they can use academic research to develop targeted environmental policies that align with the economic and social needs of BOP communities, such as incentives for sustainable resource management or support for eco-friendly local industries. This collaborative approach among multiple stakeholders ensures that environmental sustainability is framed not as a luxury, but as an integral component of economic and social well-being in BOP communities. Nevertheless, all stakeholders must remain attuned to potential conflicts between immediate economic needs and long-term environmental sustainability, striving for balanced solutions through their combined expertise and efforts.

Bringing the environmental value of sustainability to the forefront for bottom-of-the-pyramid (BoP) consumers involves creating solutions that address their immediate needs while also promoting environmental conservation and resilience. For instance, Plastics for Change, India partners with waste pickers in India to collect and recycle plastic waste, creating a circular economy for plastic materials. By reducing plastic waste and recycling it into new products, the initiative helps combat plastic pollution, reduce landfill usage, and lower environmental degradation caused by improper waste disposal. Plastics for Change turns plastic waste into a valuable resource for BoP communities, showing that recycling and upcycling can generate income while protecting the environment. This creates an economic incentive for BoP consumers to participate in environmentally sustainable practices.

Many of the products and services targeting BoP consumers offer immediate economic benefits, such as lower energy or water costs, while also conserving natural resources. This dual focus on affordability and sustainability helps BoP consumers understand the environmental value of these solutions. Solutions like clean energy and water systems demonstrate how environmental sustainability improves public health. These technologies reduce pollution, provide safe water, and promote cleaner living environments, making the benefits of sustainability tangible for BoP consumers. Programs that engage BoP consumers in sustainable practices, such as waste recycling or eco-friendly entrepreneurship, not only create jobs but also instill a sense of ownership over environmental conservation. This fosters long-term environmental stewardship. For instance, by replacing kerosene and other fossil fuels with solar energy, M-KOPA helps reduce greenhouse gas emissions and indoor air pollution. Solar energy is a clean, renewable resource, which significantly decreases the carbon footprint of households that switch from traditional fuels. The widespread use of solar power through M-KOPA's systems reduces the demand for non-renewable energy sources like kerosene, thereby lowering emissions, contributing to cleaner air, and combating climate change. M-KOPA's solar systems also reduce the need for firewood and charcoal, which many BoP households use for lighting and basic energy needs. M-KOPA's focus on energy efficiency supports broader environmental goals by encouraging sustainable development. Clean, renewable energy access is a key factor in achieving the UN's Sustainable Development Goals (SDGs), particularly those related to affordable energy, climate action, and sustainable cities. By introducing low-income consumers to solar power, M-KOPA fosters a shift toward clean energy use. Many households, after experiencing the benefits of solar energy, are more likely to adopt other eco-friendly technologies. By eliminating the need for electricity, Pureit reduces the environmental burden associated with energy consumption, particularly in areas where electricity is generated from fossil fuels. This reduces carbon emissions and promotes energy efficiency. Pureit helps households access clean water without increasing their carbon footprint, which contributes to a reduction in energy demand and greenhouse gas emissions, especially in off-grid or low-energy areas. By ensuring that contaminants are removed from drinking water, Pureit reduces the likelihood of water pollution in local ecosystems. Clean water systems help prevent pollutants from entering water bodies and harming aquatic life. Pureit targets regions where access to clean drinking water is often limited, promoting sustainable water management at the household level. Pureit's technology supports local water self-sufficiency, which is essential for the long-term sustainability of water resources in developing regions. By empowering communities to purify their own water, it reduces the need for large-scale water transportation or extraction from distant sources, which can degrade ecosystems.

By creating products and services that directly link environmental sustainability with economic savings, improved health, and community well-being, companies can help BoP consumers see the value of protecting the environment while meeting their own needs.

4.5.1 To Sum up

It is important to note that across all three dimensions of the Sustainability Pyramid—economic, social, and environmental-the goals and endeavors are inherently informed by ethical philosophy and shaped by technological advancement. Ethical considerations guide the development and implementation of sustainable practices, ensuring that initiatives are not only effective but also morally sound and culturally appropriate for BOP communities. For instance, the ethical imperative of reducing poverty informs economic sustainability efforts, while social justice principles guide social sustainability initiatives. Simultaneously, technological advancements play a crucial role in shaping how these sustainability goals are achieved. In the BOP context, appropriate and affordable technologies can accelerate progress across all dimensions of the pyramid. For example, mobile banking technologies can enhance economic sustainability, e-learning platforms can support social sustainability through education, and clean energy technologies can contribute to environmental sustainability. The integration of ethical considerations and technological innovations thus provides a robust foundation for implementing the Sustainability Pyramid in BOP markets, ensuring that sustainability efforts are both principled and practical. By creating solutions that align with the financial needs of BoP consumers, companies can make sustainability not just an environmental or ethical priority but also an economically viable one for the world's poorest communities.

In summary, the sustainability pyramid's hierarchical approach—beginning with economic benefits, followed by social improvements, and culminating in environmental sustainability—provides a practical and adaptable framework for addressing the unique challenges and opportunities at the Base of the Pyramid. By incorporating the collaborative efforts of academics, policymakers, and firms, development efforts can be aligned with the priorities and capacities of BOP communities. This tripartite approach facilitates sustainable and inclusive growth, contributing to broader goals of global development and equity.

4.6 A Case in Point—Eco Zoom's Energy Efficient CookStoves

The case of Eco Zoom's Energy Efficient Cookstoves can be discussed in alignment with the propositions offered by the sustainability pyramid. EcoZoom cookstoves are durable, efficient outdoor cooking stoves that burn with less fuel and smoke.

Proposition 1: The economic dimension of sustainability is the primary consideration in consumer decision-making toward sustainable consumption—that is, consumers will only consider the social and environmental dimensions of sustainability once the economic value of sustainability is justified.

Economic savings from reduced fuel consumption make the cookstoves highly attractive to BoP consumers. EcoZoom's energy-efficient cookstoves require significantly less wood or charcoal compared to traditional open-fire stoves or inefficient stoves. This directly reduces the household's expenditure on fuel, which is a significant cost for low-income families at the BoP. For example, a family using an EcoZoom stove can save up to 50% on their fuel costs.

Increased Household Savings: With reduced spending on fuel, families can redirect savings toward other essential needs such as education, healthcare, or small business investments, improving their overall economic well-being.

Proposition 2: The social dimension of sustainability is the secondary consideration in consumer decision-making toward sustainable consumption—that is, consumers will only consider the social dimension of sustainability once the economic value of sustainability is justified.

Traditional stoves produce a lot of smoke and harmful emissions, which contribute to respiratory diseases, especially in women and children who spend a lot of time cooking. EcoZoom's cookstoves produce far less smoke, improving indoor air quality and reducing the incidence of respiratory illnesses. This leads to better overall health outcomes and reduces healthcare costs, contributing to social sustainability. The efficiency of the cookstoves reduces the time spent gathering firewood, which is often the responsibility of women and children in rural areas. This allows children more time to attend school and provides women with more time for other productive activities, fostering gender equality and education—key components of social sustainability.

Proposition 3: The environmental dimension of sustainability is the tertiary consideration in consumer decision-making toward sustainable consumption—that is, consumers will only consider the environmental dimension of sustainability once the economic and social value of sustainability is justified.

Reduced deforestation and lower carbon emissions emerge as the ultimate environmental gains, although they may not be the main motivation for the consumers initially. By using less firewood or charcoal, EcoZoom's cookstoves contribute to reducing deforestation, which is a major environmental issue in many developing countries. This helps conserve forests, which are critical for biodiversity and play a significant role in carbon sequestration. The energy efficiency of the stoves leads to a reduction in greenhouse gas emissions, contributing to climate change mitigation. While the macro-level impacts on the environment may not be the primary driver for the consumers (who are focused on economic benefits), the micro level environmental impacts do play a pertinent role here. Traditional cookware emits harmful pollutants causing indoor air pollution, while the EcoZoom stoves reduce these emissions. These stoves, thus, improve indoor air quality and health of the users, while also contributing to a cleaner environment overall. Therefore, while the macro environment may be acting as a tertiary factor, the micro impact on the environment may have a larger role to play. EcoZoom's cookstoves offer environmental benefits at not only the macro but micro levels as well, for consumers to consider.

Proposition 4: Consumer decision-making toward sustainable consumption can be influenced by academics—that is, academics can develop new insights that inform policymaking and firm strategies as well as nurture the next generation through education and research to intervene and shape consumer decisions toward sustainable consumption more favorably.

There have been several studies identifying the impact of clean cookstoves, determinants of their adoptions, and other market factors. A report by KREA University[1] focuses on the impact of clean cooking on socio-economic, health, and environmental outcomes. A detailed understanding of these aspects may lead to informed investments in this sector, and thereby promote innovative business models. Also, policy influences can be made toward the success of such initiatives by impacting the choice

---

[1] Source: https://ifmrlead.org/wp-content/uploads/2021/12/Clean-Cookstoves-2021-Report.pdf

of business models, distribution channels, and marketing strategies which ultimately hold a significant impact on the end consumer behavior. The academic focus on clean cookstoves has found significant traction in recent times, offering actionable insights toward both policymaking and consumer behavior. For instance, Lambe et al. (2020), in their study identified three user archetypes for trying new stoves, offering insights for service design methods to better understand users, their preferences, and needs. Boudewijns et al. (2022) present evidence on factors influencing successful implementation of improved clean fuel cookstoves in low—and middle-income countries. The outcomes of the study carry implications for optimizing cleaner cooking solution implementation and thereby may impact sustainability through improving health, environment, climate, and gender equity outcomes. Gill-Wiehl et al. (2023), based on household energy surveys in Tanzania, call for household preferences to be systematically incorporated into clean cooking policy decisions. The study identifies how preference affects adoption and the need to include user preferences to meet universal clean cooking access (SDG 7).

Proposition 5: Consumer decision-making toward sustainable consumption can be influenced by policymakers—that is, policymakers can develop policies and educational campaigns to intervene and shape consumer decisions toward sustainable consumption more favorably.

Over the last few years, environmental policymaking worldwide has been giving a lower profile to the role of the State and placing increasing responsibility for many environmental issues on citizens/consumers. Individuals are increasingly perceived as agents for environmental change and their behaviors are subject to tighter scrutiny. Policymakers can play a crucial role in shaping consumer decisions toward more sustainable consumption through a combination of regulations, incentives, and educational campaigns. A few of these policy initiatives include subsidies and tax incentives on sustainable choices, regulating unsustainable practices via bans and restrictions, public information campaigns, R&D grants for green innovations, subsidized access to sustainable products, etc.

India's National Clean Cooking Program, aims to promote cleaner and more efficient biomass cookstoves for rural households. Key features of the program include subsidies to manufacturers, R&D investment support, and awareness campaigns. Kenya's Clean CookStove program features public–private partnerships and financial incentives such as the pay-as-you-go model. Also supportive regulatory frameworks have been introduced to ensure quality standards and consumer protection.

Proposition 6: Consumer decision-making toward sustainable consumption can be influenced by firms—that is, firms can develop products and strategies to attract, inform, and shape consumer decisions toward sustainable consumption more favorably.

EcoZoom employs several strategies to influence consumer decision-making. EcoZoom partners with organizations to ensure that its cookstoves are financially accessible to consumers, particularly in bottom-of-the-pyramid markets. By collaborating with organizations like the Global Alliance for Clean Cookstoves and local

NGOs, they are able to offer the stoves at reduced prices or with flexible payment options, making it easier for low-income households to afford them. By positioning the stoves as a solution to indoor pollution and respiratory diseases, EcoZoom appeals to consumers' immediate social concerns—health and well-being—especially in regions where women and children are disproportionately affected by the smoke from traditional cooking methods. Their focus on designing stoves that fit local cooking habits and materials also encourages adoption, as the stoves feel familiar and functional within specific cultural contexts. EcoZoom recognizes the pivotal role of women in household purchasing decisions, especially in communities where cooking is traditionally done by women. They work to involve women in the sales process, empowering them as entrepreneurs and community leaders who sell and promote the stoves within their local networks. EcoZoom's approach to influencing consumer decision-making for its energy-efficient cookstoves focuses on offering economic value through cost savings, improving health outcomes by reducing indoor air pollution, and making the environmental benefits secondary but essential for long-term sustainability. By using financing options, community-based marketing, and product innovation tailored to local needs, they effectively promote sustainable consumption among bottom-of-the-pyramid consumers.

Proposition 7: Consumer decision-making toward sustainable consumption can be influenced by their ethical philosophy, whereby the ethical philosophy held by consumers can be influenced by academics, policymakers, and firms to influence the consumption perception and pursuit of consumers toward economic, social, and environmental sustainability.

Consumers who are ethically motivated by environmental sustainability often prioritize products that minimize negative impacts on the environment. Studies on ethical consumption suggest that environmentally conscious consumers prefer products like energy-efficient cookstoves because they contribute to reducing deforestation and carbon emissions. Research shows that consumers are more likely to choose clean cookstoves if they are aware of their potential to reduce fuel consumption and limit emissions, which directly benefits both their immediate surroundings and the global environment (Sutar, 2022). Ethically driven consumers are also motivated by concerns for social justice and community well-being. EcoZoom, by promoting energy-efficient cookstoves that save money on fuel and reduce the burden of fuel collection (especially for women), appeals to consumers who view their purchasing decisions as a way to contribute to the well-being of their community. This is particularly relevant in low-income or rural areas, where the social impacts of fuel consumption (such as time spent gathering firewood) are felt directly.

Proposition 8: Technological advancement shapes consumer decision-making toward sustainable consumption, whereby the technological advancement leveraged by academics, policymakers, and firms can influence the consumption perception and pursuit of consumers toward economic, social, and environmental sustainability.

EcoZoom cookstoves integrate with new technologies in several key ways to enhance energy efficiency, accessibility, and user engagement. EcoZoom's cook stoves are designed using rocket stove technology, which optimizes fuel combustion by enhancing airflow and heat retention. This design helps the stoves achieve higher energy efficiency, requiring less fuel to cook the same amount of food compared to traditional stoves. EcoZoom has explored partnerships and product bundling with solar energy providers to make clean energy solutions more accessible to low-income households. For example, some pay-as-you-go (PAYG) models for solar systems in Africa bundle solar energy solutions with EcoZoom cookstoves. This integration allows households to access clean cooking and lighting technologies through affordable financing mechanisms, leveraging mobile payments and innovative distribution models. To improve stove performance and gather usage data, some companies working with clean cookstoves, including EcoZoom in partnership projects, have implemented remote monitoring technologies. Sensors embedded in cookstoves track how often the stoves are used, the amount of fuel consumed, and the reductions in emissions achieved. This data is used to assess the real-world impacts of the stoves, helping EcoZoom refine its designs and marketing strategies.

4.7 Discussion

The application of the Sustainability Pyramid to BoP markets offers a nuanced and potentially transformative approach to sustainable development. By prioritizing economic sustainability, followed by social and environmental dimensions, this framework aligns closely with the immediate needs and constraints of BoP consumers while paving the way for long-term sustainable practices. Crucially, the model recognizes that all these dimensions are informed by ethical philosophy and shaped by technological advancement. The ethical underpinnings of sustainability in BoP markets can be better understood through the lens of intergenerational and intragenerational justice. Intergenerational justice, concerning the rights of future generations, becomes particularly significant in BoP contexts where resource depletion and environmental degradation can perpetuate cycles of poverty, while intragenerational justice addresses current economic and social disparities through equitable access to resources and opportunities. These ethical considerations fundamentally inform how the pyramid's dimensions interact: while economic sustainability takes precedence, it must be pursued in ways that respect both intergenerational environmental commitments and intragenerational social equity, helping resolve tensions between immediate economic needs and longer-term sustainability goals in resource-constrained BoP environments. This ethical framework is evident in how the Sustainability Pyramid prioritizes economic stability while simultaneously recognizing the interconnected nature of social and environmental concerns, ensuring that current consumption patterns in BoP markets don't compromise future sustainability.

The pyramid's emphasis on economic sustainability as a foundation addresses the financial constraints faced by BoP consumers, as noted by Zhao et al. (2016).

This aligns with Naatu et al.'s (2024) findings that BoP consumers are more likely to engage in low-cost pro-environmental behaviors. However, the model goes beyond mere economic considerations by integrating ethical philosophies that guide decision-making across all levels of the pyramid. The role of technological advancement in shaping sustainability goals presents great opportunities in the BoP context. Technology can offer innovative solutions for sustainable development, and its implementation must be carefully considered to ensure accessibility and appropriateness for BoP communities. This necessitates a balance between leveraging cutting-edge innovations and adapting technologies to local contexts.

A significant consideration while implementing this model is ensuring that progress in economic sustainability doesn't come at the cost of social equity or environmental integrity. The ethical underpinnings of the pyramid can help navigate these potential conflicts, providing a moral compass for decision-making. Similarly, technological advancements can be leveraged to find synergies between economic growth and environmental protection. The collaborative approach involving academics, policymakers, and firms presents opportunities for comprehensive solutions but requires careful coordination. Ensuring that academic research translates effectively into policy and business practices, while respecting ethical considerations and leveraging appropriate technologies, is complex but necessary.

Furthermore, the application of this model must be sensitive to the diverse contexts within BoP markets. As Naatu et al. (2024) demonstrate, factors such as perceived severity of environmental problems and self-efficacy can vary significantly among BoP consumers. This underscores the need for localized, context-specific adaptations of the Sustainability Pyramid model, guided by local ethical norms and available technologies. It is crucial to recognize that businesses can most effectively contribute by developing a nuanced understanding of both the strengths and vulnerabilities of BoP consumers, rather than focusing on either aspect in isolation. This approach necessitates a careful examination of the similarities and differences between various BoP contexts worldwide, avoiding the assumption of a uniform character across all such markets. For instance, brand aspirations and technology receptivity can vary significantly among BoP consumers in different regions (Viswanathan & Sridharan, 2009). The reasons for poverty and the challenges faced by BoP markets differ greatly across the globe. Some BoP markets are embedded in larger climates of conflict and repression, others face severe environmental challenges, while some struggle with chronic economic stagnation (Viswanathan & Sridharan, 2009). Consequently, global businesses cannot realistically approach all these diverse BoP situations with a one-size-fits-all strategy when implementing the Sustainability Pyramid. Instead, they must tailor their approach to the specific economic, social, and environmental conditions of each BoP market, ensuring that their sustainability initiatives align with local needs and capacities at each level of the pyramid.

The Sustainability Pyramid offers a promising framework for promoting sustainable practices in BoP markets. By aligning sustainability initiatives with the economic priorities of BoP consumers, underpinned by ethical considerations and technological innovations, it creates a pathway for meaningful change. The model's emphasis on collaboration between different stakeholders, coupled with its recognition of the roles of ethics and technology, holds potential for more holistic and effective interventions.

The pyramidal approach to sustainability in BoP markets proposed in this study aligns with previous scholarly work that emphasizes the interconnectedness of economic, social, and environmental dimensions of sustainability (Giovannoni & Fabietti, 2013). While our model suggests a hierarchical prioritization, it acknowledges that these dimensions ultimately work together. The pyramid approach doesn't advocate for isolated initiatives, but for a strategic emphasis that evolves over time, recognizing that sustainable development requires an integrated effort.

In the BoP context, while economic sustainability may be initially prioritized, it's not pursued in isolation. The pyramid model provides a framework for adequately managing the tensions, trade-offs, and synergies between economic, social, and environmental aspects. This approach allows for a dynamic balance that can be adjusted based on the specific needs and circumstances of each BoP community, ensuring that all dimensions are addressed simultaneously, albeit with varying degrees of emphasis at different stages of development.

Thus, the Sustainability Pyramid serves as a guiding framework for emphasizing certain aspects at different stages of development in BoP markets, while still maintaining an integrated perspective on sustainability as a whole. This approach adapts established sustainability principles to the unique challenges and opportunities present in BoP markets.

4.7.1 Implications

This chapter demonstrates how frameworks like the Sustainability Pyramid can be contextually adapted to BoP markets. It opens avenues for developing more nuanced, context-specific sustainability frameworks. The integration of the SHIFT framework (Social influence, Habit formation, Individual self, Feelings and cognition, and Tangibility) as proposed by White et al. (2019), along with the Sustainability Pyramid offers a promising direction for future research. This integration could provide a comprehensive understanding of how psychological factors influence sustainable behaviors across the pyramid's economic, social, and environmental dimensions in BoP contexts. For instance, exploring how social influence and habit formation (from SHIFT) interact with economic priorities (from the Sustainability Pyramid) could yield insights into effective strategies for promoting sustainable practices in BoP markets. Furthermore, this study lays the groundwork for examining the dynamic interplay between sustainability dimensions, ethical philosophies, and technological advancements over time, potentially leading to more sophisticated, context-sensitive models of sustainability progression in resource-constrained environments.

The application of the Sustainability Pyramid to BoP markets, informed by ethical philosophy and shaped by technological advancement, offers significant practical implications. For effective communication of sustainability initiatives to BoP consumers, strategies should reflect the hierarchy of the pyramid while emphasizing ethical considerations and leveraging appropriate technologies. Initially, the focus should be on economically beneficial and technologically accessible sustainable practices, gradually incorporating social and environmental benefits as consumers progress through the pyramid levels. Policymakers need to create enabling environments that support this approach, potentially through tiered policy frameworks that evolve from economic development to more comprehensive sustainability initiatives, always guided by ethical principles. These policies should also encourage the development and adoption of appropriate technologies that can accelerate sustainability efforts across all dimensions. Firms operating in BoP markets should align their strategies with this ethically informed, technology-enabled pyramid approach, developing product lines or services that cater to different levels of the pyramid while adhering to strong ethical standards. This might involve creating entry-level sustainable products that offer clear economic benefits through innovative technologies, with more comprehensive sustainable offerings introduced as the market matures. Additionally, firms should consider collaborative models that leverage technological innovations and uphold ethical standards, partnering with local communities, NGOs, and governments to address challenges at each level of the pyramid effectively.

In applying the Sustainability Pyramid to BoP markets, businesses need to adopt a perspective that prioritizes a deep comprehension of the marketplace behaviors and sustainability practices of BoP consumers. This approach necessitates engaging in open, equal dialogues with these communities. By conducting marketplace research with an egalitarian and learning-oriented mindset, companies can develop sustainability initiatives that genuinely address the constraints of poverty and enhance quality of life in BoP contexts (Viswanathan & Sridhiran, 2009). This bottom-up approach allows firms to create solutions that are truly aligned with the needs and capabilities of BoP consumers at each level of the Sustainability Pyramid. Rather than imposing pre-conceived notions of sustainability, businesses can co-create strategies that resonate with local economic realities, social structures, and environmental challenges. This collaborative process ensures that sustainability efforts are not only effective but also culturally appropriate and economically viable for BoP markets.

4.7.2 Limitations and Future Research

This study's conceptual nature limits its empirical validation in real-world BoP contexts. Future research can conduct field studies to test the Sustainability Pyramid's effectiveness across diverse BoP markets, considering varying cultural, economic, and environmental factors. The study's broad focus on BoP markets may overlook important regional variations. Future research could explore how the model's application differs across urban and rural BoP segments, or in markets at different stages of development. Another limitation is the potential oversimplification of transitions between pyramid levels. Further research is needed to understand the nuanced processes by which BoP consumers and communities progress from economic to social and environmental sustainability concerns. Future studies could also investigate the role of local knowledge and traditional practices in implementing the Sustainability Pyramid, exploring how these can be integrated with modern approaches to create more effective and culturally appropriate sustainability strategies. Lastly, there's a need for interdisciplinary research that integrates perspectives from various fields such as development economics, sociology, and environmental science. This could provide a more holistic understanding of sustainable development in BoP markets and help refine the Sustainability Pyramid model for greater applicability across diverse contexts.

References

Achrol, R., & Kotler, P. (2016). Marketing's lost frontier: The poor. *Markets, Globalization & Development Review, 1*(1).

Boudewijns, E. A., Truchhi, M., Ven Der Kleij, R. M., Vermont, D., Hoffman, C. M., & Chavannes, N. H. (2022). Facilitators and barriers to the implementation of improved solid fuel cookstoves and clean fuels in low-income and middle-income countries: An umbrella review. *The Lancet Planetary Health, 6*(7).

Costanza, R., & Patten, B. C. (1995). Defining and predicting sustainability. *Ecological Economics, 15*, 193–196. https://doi.org/10.1016/0921-8009(95)00048-8

Demssie, Y. N., Wesselink, R., Biemans, H. J. A., & Mulder, M. (2019). Think outside the European box: Identifying sustainability competencies for a base of the pyramid context. *Journal of Cleaner Production, 221*, 828–838. https://doi.org/10.1016/j.jclepro.2019.02.255

Gill-Wiehl, A., Sievers, S., Katikiro, R., & Kammen, D. M. (2023). Evaluation of the preference for and viability of clean cookstove adoption in rural Tanzania. *Energy, Sustainability & Society, 13*(42).

Giovannoni, E., & Fabietti, G. (2013). What is sustainability? A review of its concept and applications. In C. Busco et al. (Eds.), *Integrated Reporting*. Springer International Publishing. https://doi.org/10.1007/978-3-319-02168-3_2

Gray, R. (2010). Is accounting for sustainability actually accounting for sustainability... and how would we know? An exploration of narratives of organisations and the planet. *Accounting, Organizations & Society, 35*(1), 47–62.

Hammond, A. L., Kramer, W. J., Katz, R. S., Tran, J. T., & Walker, C. (2007). The next 4 billion. Innovations: Technology. *Governance, Globalization, 2*(1–2), 147–158.

Hahn, R. (2009). The ethical rational of the business of the poor—integrating the concepts bottom of the Pyramid, sustainable development & corporate citizenship. *Journal of Business Ethics, 84*, 313–324.

Kelleci, A., & Yildiz, O. (2021). A guiding framework for levels of sustainability in marketing. *Sustainability, 13*. https://doi.org/10.3390/su13041644

Lambe, F., Ran, Y., Kwamboka, E. et al. (2020). Opening the black pot: A service design-driven approach to understanding the use of cleaner cookstoves in peri-urban Kenya. *Energy Research & Social Science*, 70.

Lim, W. M. (2022). The sustainability pyramid: A hierarchical approach to greater sustainability and the United Nations sustainable development goals with implications for marketing theory, practice, and public policy. *Australasian Marketing Journal, 30*(2), 142–150.

McDonagh, P., & Prothero, A. (2014). Sustainability marketing research: Past, present and future. *Journal of Marketing Management, 30*(11–12), 1186–1219.

Moore, J. E., Mascarenhas, A., Bain, J., & Straus, S. E. (2017). Developing a comprehensive definition of sustainability. *Implementation Science, 12*. https://doi.org/10.1186/s13012-017-0637-1

Naatu, F., Alon, I., Amoako, G. K., & Mornah, D. (2024). Pro-environmental values' and consumer behavior in base of the pyramid market: Ghana. *Competitiveness Review: An International Business Journal, 34*(4), 807–831.

Ngosso, T. (2013). The right to development of developing countries: An argument against environmental protection? *Public Reason, 5*, 3–21.

Prahalad, C. K., & Hart, S. L. (2002). Strategy+ business. *The Fortune at the Bottom of the Pyramid, 26*(1), 2–14.

Prahalad, C. K. (2005). The bottom of the pyramid–eradicating poverty through profits.

Rahman, S. A., Amran, A., Ahmad, N. H., & Taghizadeh, S. K. (2015). Supporting entrepreneurial business success at the base of pyramid through entrepreneurial competencies. *Management Decision, 53*(6), 1203–1223.

Rieckmann, M. (2013). The global perspective of education for sustainable development: A European-Latin American study about key competencies for thinking and acting in the world society. *Environmental Education Research, 19*(2), 257–258. https://doi.org/10.1080/13504622.2012.697547

Scheyvens, R., Banks, G., & Hughes, E. (2016). The private sector and the SDGs: The need to move beyond 'business as usual.' *Sustainable Development, 24*(6), 371–382.

Sutar. (2022). Energy efficiency, emissions and adoption of biomass cookstoves. In Shahzad et al. (Eds.), *Alternative energies and efficiency evaluation*. IntechOpen. https://doi.org/10.5772/intechopen.101886

Tomšič, N., Bojnec, Š, & Simčič, B. (2015). Corporate sustainability and economic performance in small and medium sized enterprises. *Journal of Cleaner Production, 108*, 603–612. https://doi.org/10.1016/j.jclepro.2015.08.106

United Nations (UN). (1972). Report of the United Nations conference on the human environment. http://www.un-documents.net/aconf48-14r1.pdf

United Nations. (2016). The sustainable development goals report 2016. https://unstats.un.org/sdgs/report/2016/The%20Sustainable%20Development%20Goals%20Report%202016.pdf

United Nations Department of Economic and Social Affairs. (2021). The sustainable development goals report 2021. https://unstats.un.org/sdgs/report/2021

Venugopal, S., & Chakrabarti, R. (2022). How subsistence communities reconfigure livelihood systems in response to climate change: A coupled-systems perspective. *Journal of Macromarketing, 42*(2), 292–307.

Venugopal, S., Mohapatra, S., & Aiyar, A. (2024). Crafting climate solutions for nature-dependent subsistence communities: A dignity-centric view. *Journal of Macromarketing, 02761467241238177*.

Viswanathan, M., & Rosa, J. A. (2010). Understanding subsistence marketplaces: Toward sustainable consumption and commerce for a better world. *Journal of Business Research, 63*(6), 535–537.

Viswanathan, M., & Sridharan, S. (2009). From subsistence marketplaces to sustainable marketplaces: A bottom-up perspective on the role of business in poverty alleviation. *Ivey Business Journal, 73*(2), 1–15.

Vohra, A., & Vishnoi, P. (2022). Environmental consciousness and sustainability for urban BoP consumers in India. In *Marketing to the Poor* (pp. 72–93). Routledge India.

Voola, R., Bosangit, C., Polonsky, M., Rosenbloom, A., & Goswami, P. (2024). Embedding the United Nations Sustainable Development Goals (SDGs) in marketing education. *Journal of Marketing Education, 46*(2), 79–83.

Voola, R., & Voola, A. P. (2019). Developing sustainable strategies at the base of the pyramid. *Ethics, Social Responsibility and Sustainability in Marketing* (263–276).

White, K., Habib, R., & Hardisty, D. J. (2019). How to SHIFT consumer behaviors to be more sustainable: A literature review and guiding framework. *Journal of Marketing, 83*(3), 22–49.
White, M. A. (2013). Sustainability: I know it when I see it. *Ecological Economics, 86*, 213–217. https://doi.org/10.1016/j.ecolecon.2012.12.020
Zhao, G., Cavusgil, E., & Zhao, Y. (2016). A protection motivation explanation of base-of-pyramid consumers' environmental sustainability. *Journal of Environmental Psychology, 45*, 116–126.

Prama Vishnoi is a faculty member at Delhi University, where she has been teaching for over a decade. She earned her PhD from the prestigious Faculty of Management Studies (FMS), Delhi, with a focus on Bottom of the Pyramid (BOP) markets. With eleven years of academic experience, she has contributed significantly to the field through her teaching and research, exploring how businesses can create value while addressing the needs of economically disadvantaged communities.

Anupama Vohra is a former Professor at the Faculty of Management Studies, University of Delhi. She brings in extensive teaching experience of more than three decades in Marketing Management, including Sales and Distribution and B2B Marketing; and has also contributed to executive training programs for various organizations. With a focus on Consumer Behavior, her research has been widely published in leading national and international journals. She co-authored the book MM: A South-Asian Perspective, by Cengage Publications, and has successfully led several research projects funded by organizations such as WHO, ICSSR, UGC, and Delhi University.

Neha Bhardwaj currently holds the position of Assistant Professor at the Faculty of Management Studies (FMS), University of Delhi. Previously, she served as a faculty member at the Indian Institute of Management (IIM) Rohtak. Her research primarily centers on consumer behavior, particularly examining how emerging technologies influence consumer choices and behaviors. Through her research, she has delved into the impact of technology on subsistence markets, specifically within the realm of marketing. Her research also investigates the role of technology in the lives of non-BOP consumers.

Chapter 5
Bridging Vulnerability and Sustainability: Collective Climate Change Adaptation Efforts by Indian Communities and Businesses

Falaq Mubashir, Tahir Ahmad Wani, and Ramendra Singh

Abstract In developing countries like India climate change poses a huge threat, as the dependence of natural resources for survival is quite high. In this chapter, we examine the inter-connectedness of climate change and poverty targeting bottom-of-the-pyramid communities. These communities are often required to deal with both natural dangers as well as limited availability of resources making their life severely difficult. The chapter further evaluates numerous coping strategies, including technological solutions, agricultural innovations, and migration, changing cropping patterns, borrowing and financial support, livestock management, utilizing natural resources as a coping mechanism. Additionally, the chapter highlights the correlation between gender and climate change. It discusses the difficult experiences of women and the specific challenges they face caused by climate change and highlights how they struggle more than men in managing changing environmental conditions. Furthermore, the study puts forward some significant gender-based coping mechanisms for battling climate change that can uplift women and reduce their vulnerability including providing them education, promoting gender equality, infrastructure development, giving them resources and property rights and more. Lastly, the chapter focuses on the role of business organizations in managing climate change and further sheds light on how they gain resilience by indulging in sustainable practices and technology. Consequently, highlighting the social and financial challenges businesses face when adopting climate-friendly procedures due to the possibility of unwanted outcomes caused by extensive interventions. Thus, by examining the joint effort of Indian companies and communities in response to climate change this

F. Mubashir (✉) · T. A. Wani
HSS&M, National Institute of Technology Srinagar, Srinagar, India
e-mail: falaqsahaf@gmail.com

T. A. Wani
e-mail: tahirwani@nitsri.ac.in

R. Singh
IIM, Calcutta, India

R. Singh and T. A. Wani (eds.), *Marketing to the Base of the Pyramid*,
https://doi.org/10.1007/978-981-96-7105-2_5

chapter underscores the importance of integrated, multi-stakeholder approaches to building climate resilience.

Keywords Sustainability · Base of Pyramid · Climate Change · Communities

5.1 Introduction

Climate change is recently considered to be one of the greatest problems of the contemporary world, especially in relation to those who are in the lower rung of the economic ladder or BOP. The findings of climate change studies reveal that both intensity as well as the extent of climate change and its effects jeopardize even the most basic poverty eradication efforts in the least developed countries (Adger et al., 2003; IPCC, 2001). According to the Millennium Ecosystem Assessment (2005), climate change plays a central role in the degradation of ecosystem services which, in turn, results in aggravation of poverty especially to people who depend on natural resources for their livelihoods.

In countries in South Asia and Sub-Saharan Africa, climate change could push millions into poverty, which would have far-reaching consequences for food security, health, and economic status (Stern, 2006; World Bank Report, 2009). The poorest populations in these areas are more vulnerable as they engage in climate-sensitive activities such as agriculture and fishing, making them more susceptible to shifts in weather patterns, rising sea levels, and the increased frequency of extreme climatic events (Nanda, 2009; Simms et al., 2004).

Moreover, BOP communities have to face tough challenges due to their limited access to financial resources, technology, and social safety nets, which ultimately makes adaptation to climate change particularly difficult (Adger et al., 2003; Smith, 2000). Despite being the least contributor to pollution and global emissions, these communities are often located in regions more prone to climate-induced hazards, such as droughts, floods, and land degradation (Ebi et al., 2007). Thus, in these vulnerable communities, climate change badly impacts and threatens their sustenance and overall health (Davidson et al., 2003; O'Brien et al., 2007).

Since climate change and poverty are inter-connected, it is crucial to devise certain adaptation strategies like providing proper irrigation facilities, making available drought resistant, heat resistant seeds at nominal price or facilitating technology-based solutions that also take into consideration the socioeconomic status of the BOP. In India, for example, the effects of climate change which include fluctuations in rainfall pattern, high temperatures, and extreme weather conditions are increasingly affecting BOP communities adversely. Besides, these communities are mostly involved in subsistence farming and informal forms of employment, thus making them highly vulnerable to rising sea levels, shifting precipitation patterns, and frequent disasters such as floods, droughts, and cyclones, all of which can negatively impact their economic growth and social well-being (Karl et al., 2009; Stern, 2006). Moreover, the climate induced adversities multiply the existing inequalities, making it even more difficult for those who have limited to no access to financial resources, education, or healthcare to adapt (Schipper & Lisa, 2007). Additionally, the other prevalent underlining issue associated with climate change is its adverse effects on women. Although the impact of climate change is generalized however, the severity of this impact is more on women compared to men (Arora-Jonsson, 2011; Stone et al., 2022).

Now, India faces 2 types of challenges: reducing poverty while simultaneously taking into consideration the threats and challenges posed by impacts of climate change. To meet this challenge, it is crucial to fathom the specific vulnerabilities of the BOP segment. This chapter critically analyzes the existing literature on how global climate change has adversely impacted India's poor and how it have posed economic, social, and financial challenges to them (Nanda, 2009; Nath & Behera, 2011).

5.2 Understanding Climate Change Vulnerabilities

It is highly crucial for countries to understand the impact of climate change, as vulnerability to it differs from one region to another and from one sector to another (Nanda, 2009). Climatic stresses resulting from climate change have been observed to affect ecologically sensitive regions more severely (Behera & Vaswani, 2007). This is especially true for poor and vulnerable socio-economic groups, whose livelihood resources largely depend on natural resources. In sectors like agriculture, vulnerability varies because communities exposed to multiple stresses employ various coping mechanisms (Acosta-Michlik & Espaldon, 2008).

In this context, the differences in how sectors adapt to vulnerability or respond or fail to respond to vulnerability within a particular region are influenced by factors such as ownership of assets (land, livestock, irrigation equipment), education levels, and social networks (Acosta-Michlik et al., 2008). Actual adaptation to vulnerability is also significantly determined by social, economic, and institutional factors such as land rights systems, irrigation availability, and access to affordable credit (Saldana-Zorrilla, 2008).

Recent research body highlights the need for localized identification of vulnerability to reflect the significant social, economic, and natural differences of communities (Ali & Kapoor, 2008; Behera & Vaswani, 2007; Byg & Salick, 2009; Deressa et al., 2008; Keskitalo & Kulyasova, 2009; Kumar & Mohanasundari, 2024; Painter et al., 2024; Smith & Diedrich, 2024). Local assessments help us understand how and why communities react differently to similar types of environmental stress.

Although a vast body of literature already exists on the effects of climate change, much of this work has focused on developed countries, particularly the United States (Adams et al., 1999; Easterling et al., 1993; Mendelsohn et al., 1994; Mendelsohn & Dinar, 2003). While these studies have improved our understanding of the consequences of climate change in developed nations, they provide little insight into what may be expected in large and diverse developing countries like India, which are highly sensitive to climate change yet remain largely under-researched.

Recent studies have begun efforts to evaluate the effects of climate change in developing countries, especially in Africa and parts of South Asia. In developing countries such as India, a large segment of the population relies on agriculture for their livelihood but lacks access to quality climate information, better agricultural inputs, non-farm employment, and support structures (Nanda, 2009). Climate change poses a significant threat to their lives since they depend heavily on climate-sensitive products. Current climatic changes may lead to a decline in the availability of natural resources that rural households use for consumption or marketing.

The Intergovernmental Panel on Climate Change (IPCC) has identified key areas vulnerable to the impacts of rising global temperatures: freshwater, life support systems, food, coastlines, and health (IPCC, 2007a). In the Indian context, food and water security are of utmost significance. Flood risks may increase during the wet season, while water availability during the dry season may decrease due to the melting of glaciers in the Himalayas, which are crucial to millions of people (Stern, 2006). Additionally, climate variations such as cyclones and changes in monsoon patterns present enormous challenges for rain-fed agriculture, which most of India's population depends on (Kumar et al., 1999). Agricultural productivity can be directly influenced by temperature and rainfall (IPCC, 2001). It can also be affected indirectly by factors such as soil condition, pests, and diseases. With reduced water supply and increased susceptibility to pest and disease attacks, poor farmers are likely to be the most adversely affected. In many areas of India, crop development is already close to its optimal temperature limits, and even slight changes in climate can cause significant yield losses. Thus, India is highly vulnerable to challenges posed by climate changes. However, the climate change itself as well as vulnerability of BOP communities is noted and continuous work on building coping mechanisms against the adversities caused by it is being carried out by individuals, government agencies as well as business organizations. This leads us to the next part of this chapter which is to discuss adaptation strategies implemented by BOP communities to survive climate change.

5.3 Adaptation Strategies in the Context of Poverty and Rural Livelihoods

Several researchers have pointed out measures that rural communities sensitive to climate change can use (Adger et al., 2003; Hulme & Shepherd, 2003). Climate adaptation requires changes in societal and economic activities, as well as improved coping options for actual or potential climatic adversities (Gallopín, 2006). According to Adger (2003), social capital is usually regarded as a societal asset that facilitates collective actions. Therefore, the capacity and scale of adaptation depend on the extent of impact on people and natural systems. Here, vulnerability is defined as the capacity to be adversely affected by exposure to stresses resulting from environmental and social change, along with the inability to cope with these challenges (Adger, 2006).

Poverty is both a condition and a cause of vulnerability; hence reducing vulnerability involves directly addressing poverty. The Multidimensional Poverty Index includes not only income but also other essential requirements for a wholesome life, such as health, nutritional status, education, standard of living, agency, and social connections in India (Sen, 1999). Building on people's basic skills improves their capacity to describe, question, resist, and transform new actions, which is vital for adaptation.

Forced to rely on available resources, traditional Indian communities depend on weather-sensitive physical resources such as forests, the sea, and livestock through collective adaptation processes (Agrawal, 2001; Alexander et al., 2010). However, these adaptation strategies are often formulated as reactive or event-driven strategies rather than planned ones (Ellis, 1999; Nuorteva et al., 2010). They are usually eclectic and responsive rather than proactive, which contrasts with strategic approaches like the Adaptive Behavioral Model (ABM) (Smith, 2000). These strategies are more often employed in socio-economic sectors characterized by low capital intensities (Sohngen & Mendelsohn, 1998). Some community-managed adaptation practices in India include crop diversification, traditional water management, and disaster risk management (IPCC, 2007b).

Due to the increased effects of climate change on the continent, rural households in Africa, specifically in South Africa, have adopted various strategies to cope with the challenges caused by climate change. These coping strategies range from conventional to contemporary approaches. Tropical communities have relied on indigenous knowledge systems and natural resources, including non-timber forest products, to mitigate the impacts of climatic variation (Locatelli et al., 2010; Pramova et al., 2012; Shackleton et al., 2008). For example, in Limpopo province of South Africa, farmers have adopted new strategies such as switching from maize to drought-resistant crops like sorghum, alongside irrigation practices, which are crucial for their survival (Chigavazira, 2012).

Some adaptive measures include changing planting dates, improving irrigation methods, and harvesting water. Furthermore, some households have described risk management strategies, such as planting trees and selling off livestock (Bryan et al., 2010). However, there are variations in how people and communities respond to similar climatic impacts, even within the same region (Bryan et al., 2009).

The factors influencing the choice of adaptation strategies are rather conditioned. Ten percent of the variance can be attributed to socio-economic factors such as credit, resource access, extension services, and/or perception of risk (Adeniji-Oloukoi et al., 2013; Berman et al., 2013). Similarly, access to markets determines the type of crop cultivated; for example, millet is grown in some areas due to its drought resistance and its ready market (Bryan et al., 2009).

However, other factors, including limited access to credit, technical know-how, and markets, remain significant hindrances to adaptation processes. For instance, food relief programs have, in some ways, hindered long-term community adaptation measures such as tree planting (Bryan et al., 2010). Additionally, cultural preference given to specific crops, like maize rather than sorghum, and lack of knowledge regarding crop management also reduces the adaptive capability of these communities (Quinn et al., 2011).

- **Migration as coping mechanism:** Climate-induced adversities and challenges have made migration as one of the coping mechanism in India, with specific emphasis on rural agrarian-based populations. As a last resort, due to the unpredictable rainfall, reduction in crop yields and implications of food insecurity, migration is perceived by farmers as a strategy that will maximize income sources, thus, improving household flexibility. Several studies have established that climate-induced migration in India is a response to livelihood vulnerability, where households in rural areas migrate permanently or seasonally to urban areas or other regions for income (Black et al., 2011; Craven, 2015). For instance, in India, migration is higher during the Rabi season due to fewer work opportunities and low-value crop yields, while in the Kharif season, migration is lower because of the demand for farm labor for the production of rice and other crops considered of high-value (Sakdapolrak et al., 2014). Also, the money sent by families that migrate helps in increasing the households' financial resilience and coping capacity in order to invest in better techniques, technologies, and policies on agriculture and climate change adaptation (Abu et al., 2014; Afifi et al., 2014). Migration thus not only aids in maintaining consumption smoothly but also assists in availing agricultural extension services and modern approaches to farming, which in turn improves the vulnerable population left behind (Etzold et al., 2014). However, as a coping strategy, migration may also act to strip rural areas off their labor and skills base, which may further be detrimental for the rural economy and cause overall decrease in adaptive capacity (Feng et al., 2010). Hence, continued migration is an effective strategy for adaptation to climate change impacts, though it has to be backed by sound policies which would not harm the rural livelihoods in the long run.
- **Changing Cropping Patterns:** By switching to crops that are drought-resistant or comparatively less water-intensive farmers have eventually been able to cope with changes in climate. This strategy has been useful in regions prone to cyclical droughts, since it enables farmers to continue farming even during unfavorable climate conditions, thereby preventing negative impacts on their productivity. Similarly study conducted by Legide et al., (2024) examines the ways in which smallholder farmers in the Eastern Cape Province of South Africa have established a nuanced system for gathering, evaluating, and forecasting temperature and rainfall data. This knowledge empowered them to ascertain optimal cropping patterns and planting schedules (which function as a coping mechanism in response to climate challenges). These farmers depended on their indigenous weather forecasting abilities; they incorporate indicators like plant phenology, animal behavior, and moon phases to anticipate the timing, intensity, and distribution of rainfall. However, this approach to adapting cropping patterns enabled farmers to make well-informed decisions regarding crop selection and timing. Consequently, it enhances their resilience to climate variability and change. Henceforth it can be

seen that, farmers consider cropping pattern change, improved irrigation, less water consuming crop varieties as the strategies to cope with climate change with least investment (Singh, 2020).

- **Borrowing and Financial Support:** Another way to cope with constraints is through taking loans from kinship networks or receiving complementary government support. However, this is less often practiced than food rationing or changes in cropping techniques. This assistance helps families cover small expenses during the dry seasons in an effort to support the agriculture sector. Moreover, India has consistently assumed a prominent position in underscoring the issue of climate finance at the UNFCCC (United Nations Framework Convention on Climate Change) and other multilateral platforms. Its initiatives have effectively contested the exaggerated assertions made by developed nations regarding their progress toward achieving climate finance targets; this revelation exposes the reality that the actual funds mobilized are considerably lower. Furthermore, India has been at the vanguard of advocating for a clear and precise definition of climate finance, emphasizing the necessity for substantial scale, a broader scope, and expedited delivery. It has steadfastly asserted that climate finance must be new and additional to existing overseas development assistance (ODA), primarily provided in the form of grants rather than loans and should strike a balance between mitigation and adaptation efforts. However, the complexities surrounding climate finance continue to pose challenges for all stakeholders involved (Press Information Bureau, Government of India, 2022).
- **Livestock Management:** Livestock is a major resource for managing coping strategies in rural households. Hence, livestock management assumes a vital role in addressing the challenges posed by climate change, functioning as both a resilient and adaptive element within agricultural systems. Ownership of livestock, particularly among smallholder farmers, serves as a strategy for mitigating climate-related risks; it offers income diversification and operates as capital assets (Herrero et al., 2016; Weindl et al., 2015). Most families sell their livestock during drought periods to maintain financial stability. Additionally, some families, especially those with small-scale farming and a number of goats and native cattle, often use local cattle that do not require much water and fodder. For instance, goats can feed freely in the surrounding forests, helping families survive during hardships (Kvanneid, 2021; Yadav et al., 2008). However, unlike crops, livestock tends to exhibit greater resilience to significant climate variability, thereby providing a buffer against climate-induced threats such as droughts, floods, and heat waves (Sejian et al., 2015). Although effective adaptation strategies encompass a range of farm management adjustments, technological innovations and income-related responses, such as diversifying livestock species or incorporating mechanization and cooling systems, these measures are crucial for alleviating the impacts of extreme weather events (Escarcha et al., 2018; Henry et al., 2018). The multifaceted approach helps to combat key climate hazards, including inter-annual forage variability, which is essential for ensuring food security and sustainable livelihoods (Salman et al., 2019). Furthermore, livestock plays a crucial role in fostering long-term resilience: it underpins the socioeconomic stability of

vulnerable communities. This highlights its significance in constructing climate-resilient agricultural systems (Herrero et al., 2020). However, one must consider the complexities involved in such systems, because they require careful planning and execution. Although these benefits are apparent, the challenges cannot be overlooked.

- **Utilizing Natural Resources:** Utilizing natural resources includes the use of firewood and foliage by rural communities to meet their needs. One method for tackling energy poverty and utilizing natural resources involves the adoption of more efficient cooking devices and alternative fuels. Feedback from focus group discussions (FGDs) in the study conducted by Njenga et al., (2024) indicated that both refugees and women from host communities were receptive to experimenting with techniques aimed at reducing fuel consumption. Cooking energy innovations must consider various factors: socio-cultural preferences, compatibility with local cooking practices, affordability, fuel efficiency, a reduction in household air pollution, and environmental sustainability. During the study, a refugee woman proudly demonstrated her hand-crafted three-burner clay cookstove, which was designed to accommodate different pot sizes and utilized charcoal, briquettes, and firewood. The combination of briquettes with locally improved cooking devices appears promising for broader adoption. However, there is a crucial need to link briquette producers with charcoal manufacturers who utilize the invasive Prosopis Juliflora, which has the potential to supply charcoal dust as the species continues to proliferate (FAO, 2018). Furthermore, supporting these enterprises with facilities for briquette production could significantly enhance their growth. However, the use of these resources is limited due to government policies.
- **Role of technology:** The strategies focusing on climate change coping mechanisms are not just confined to above mentioned strategies, but there is a huge role technology can play. For example, India's climate technology startup landscape is experiencing rapid growth, characterized by substantial investments in energy transition, clean mobility, and sustainable agriculture. However, despite a global decline in climate tech funding during 2023, India continues to prioritize the expansion of renewable energy. This commitment is exemplified by ambitious targets such as achieving 500 GW of non-fossil fuel capacity by 2030. Solar PV (photovoltaics), a prominent sub-sector, has undergone impressive growth, bolstered by regulatory measures, including production-linked incentives. Although India is capitalizing on renewable energy for both mitigation and adaptation, it is also embracing innovations like distributed energy systems and hybrid projects. To fulfil global climate objectives, increased investments and technological advancements in renewable and decentralized solutions are essential (because they pave the way for a more sustainable future). Further, investment in key climate technologies can significantly contribute to both decarbonization and climate adaptation efforts in India (Krishna Chaitanya, 2023). Investment in crucial climate technologies can, indeed, play a significant role in both decarbonization and adaptation efforts in India (World Bank Group, 2023). Renewable energy-based technologies, such as solar, wind and hybrid projects, have the potential to diminish reliance on fossil fuels, thereby supporting sustainable energy

access in remote regions. Electric vehicles (EVs) are another important aspect; by promoting their adoption, greenhouse gas emissions from transportation can be reduced, which in turn improves air quality in urban areas. Furthermore, climate-smart agriculture practices including precision farming, drought-resistant crops, and efficient irrigation systems can enhance food security and resilience to climate variability, which is critical for sustaining agricultural output. Green hydrogen is an emerging fuel source that offers a cleaner energy alternative for both industrial processes and heavy transportation sectors. Additionally, carbon capture technologies, designed to capture and store carbon emissions, can assist industries in achieving their decarbonization targets while mitigating adverse environmental impacts. Digital technologies, such as AI, IoT and blockchain, play a vital role in facilitating efficient resource management, predictive analytics, and enhanced climate risk assessments. These innovations are essential for formulating effective adaptation and mitigation strategies, thus illustrating the multifaceted approach needed to address climate challenges. However, this complex interplay between technology and climate action underscores the necessity for a concerted effort to integrate these solutions into broader environmental policies. The substantial initial expense (along with the necessary infrastructure) associated with the adoption of emerging technologies can, however, restrict their accessibility for rural communities. This limitation is significant because it hampers the potential for these areas to benefit from advancements. Although some may argue that innovation can reach all corners of society, the reality is that financial constraints remain a critical barrier. Therefore, it becomes imperative to address these challenges to ensure equitable access.

5.4 Gender and Climate Change

Research reveals that climate change has a greater impact on the lives of women and girls, especially in areas affected by events such as droughts. In such areas, school dropout rates for girls increase as they are forced to drop out to help with domestic work or contribute to the little income their parents or families may have (Brody et al., 2008). This has been observed in Malawi and in the Thar Desert of Rajasthan, India, where most women are still illiterate, and thus the cycle of poverty and gender discrimination continues (Sandhya & Dashora, 2003). Climate change, for instance, results in the displacement of men in search of employment, leaving women with many chores, which forces them to receive low wages or be subjected to severe abuse by employers, such as in the brick kilns, where they sometimes work without pay and are harassed (Alfthan et al., 2011). Additionally, natural resource-based climate change related disasters increase the vulnerability to human trafficking, particularly for women and girls, for labor and sexual exploitation (Curtol et al., 2004; Alfthan et al., 2011). They are at higher risk of becoming involved in sexual practices, which increases their vulnerability to becoming pregnant, contracting STDs, or facing sexual coercion (Alston, 2014).

5.5 Gender-Based Coping Mechanisms for Combating Climate Change: The Experience of Women

Hence, women, especially those in rural areas, are the most affected gender by the current effects of climate change. Nevertheless, their roles in the processes of mitigation and adaptation are unique and very important.

Various strategies that focus on empowering women and reducing their vulnerability to climate change have been identified:

Poverty Elimination and Economic Dignity: The issue of poverty increases women's vulnerability to climate change. Creating employment and business opportunities for women, implementing credit policies, and promoting savings and insurance in changing situations can help women adapt. According to research conducted by Grown et al. (2006), Prowse et al. (2009), and Hill (2011), women with agency are better equipped to foster family protection, environmental protection, and climate change mitigation.

Infrastructure Development: Access to affordable energy, a well-developed transport network, and reliable water supply can help reduce the burden on women and enhance their effectiveness in adapting to current and future challenges. By decreasing reliance on fossil-based energy sources, increased access to basic commodities allows women to better combat the effects of climate change (Aguilar, 2009; Hill, 2011; Nwoke & Ibe, 2014).

Defending Resources and Property Rights: Ensuring parity in land ownership for women and stable tenures to resources is crucial for enhancing productivity and tackling shocks from climate change. Women can effectively manage larger tracts of land and other essential resources required to support their families, which also facilitates easier access to loans or other financial instruments that are helpful during economic shocks (Rodgers & Menon, 2013; WOCAN, 2014).

Promoting Gender Equality: When women of reproductive age are denied their rights to control their reproductive health and fertility, sustained unequal access to resources and opportunities results in climate injustice for them. The allocation of balanced power in policy frameworks, as well as in decision-making processes, is vital for climate change adaptation. It is argued that engaging women in opportunities equal to those of men enables them to cope with the risks posed by climate change (Haigh & Vallely, 2010; Solar, 2010).

Education and Information Dissemination: Fighting climate change requires that women receive education to reduce their vulnerability to climatic changes. Adult literacy, vocational training, and other skills development initiatives aimed at helping women adapt to new changes enhance their capacity. Moreover, there is evidence that timely release of early warnings and disaster preparedness information strengthens them (Lutz et al., 2014; World Bank, 2004).

Diversification of Livelihoods: To decrease women's reliance on climate-sensitive activities such as farming and herding, it is recommended that more diverse sources of income be promoted among women. Rural resilience is therefore determined by sustainable resources and climate-smart agricultural practices, as highlighted by Ajani et al. (2013) and Lambrou and Piana (2006).

Establishment of Policy Frameworks and Interagency Coordination: According to the study, climate change adaptation entails the integration of women's issues in policies concerning land, water, energy, and agriculture. Synergies between agencies active in the areas of poverty, women, and climate change can result in co-benefits for climate change mitigation and women's adaptation (Lambrou & Nelson, 2010; CDKN, 2011).

5.6 Emerging Trends in Indian Business Context Against Climate Change

Businesses play a significant role in enabling BOP communities to adapt to climate change; this is achieved by tackling their distinct vulnerabilities with innovative and inclusive strategies. These endeavors not only minimize the vulnerability, exposure, and sensitivity of BoP communities toward climate risks, but also boost their adaptive capacity. By building resilience through sustainable business practices and affordable technologies, businesses can make a significant impact. Although challenges exist, the potential benefits are substantial because they empower communities to thrive in the face of adversity.

One of the methods to achieve this is the integration of BoP (Bottom of the Pyramid) communities into value chains, serving as producers or entrepreneurs; this not only gives them stable income opportunities but also grants them access to vital climate knowledge. Companies such as Jain Irrigation have excelled in this domain, offering affordable drip irrigation systems and drought-resistant seeds to tackle agricultural challenges resulting from erratic weather patterns (*Endeva,* 2015*)*. Moreover, Jain Irrigation provides purchase guarantees at pre-agreed prices, reducing financial risks for smallholder farmers; however, this also creates a reliable market for its products. Such initiatives exemplify a win–win situation: farmers acquire financial security and resources to combat climate hazards, while businesses cultivate robust supply chains. Further, companies have started to work on the development of adaption strategies such as heat stress resistance in crops and animals and + diversifying crop rotations (Beck, 2024). Another critical strategy is to minimize the vulnerability of these communities by minimizing the exposure to climate risks. This can be done by introducing technology-based solutions by businesses for these communities to survive climate change induced adversities. For example, making weather prediction apps or early warning systems available for these communities can enable them to take proactive measures, such as adjusting their planting schedules or harvesting according to the weather conditions. Weather insurance companies, like Munich Re, further propagate this approach by sharing real-time information offering weather-indexed insurance, thereby enabling communities to safeguard their assets and livelihoods against extreme events.

Businesses (in many ways) contribute to enhancing the adaptive capacity of individuals at the Base of the Pyramid (BoP) by providing them with essential knowledge and resources. For instance, Unilever, through its collaboration with the Rainforest Alliance, offers training to farmers for sustainable agricultural practices, erosion control, and biodiversity conservation. These initiatives not only bolster the climate resilience of farmers; however, they also guarantee sustainable sourcing for the company. This demonstrates how businesses are propagating climate adaptation that coincides with corporate strategy. Similarly, affordable irrigation systems or drought-resistant seeds sold by companies at a nominal cost can further encourage sustained use, thereby fostering both immediate and long-term resilience. It indicates that deeply discounted pricing tends to be more effective in promoting long-term adoption and usage of climate-resilient technologies (Jones Christensen et al., 2015). Furthermore, companies like GreenJams (which offers Agrocrete innovation) develop cost-effective, carbon-negative construction materials derived from agricultural residues. Such advancements make sustainable building solutions more accessible to low-income groups. Similarly, climate tech firms, such as Ecozen, provide solar-powered solutions that aim to enhance agricultural yields, thereby directly improving the livelihoods of small farmers. These businesses not only address the immediate impacts of climate change for BoP communities; they also promote long-term adaptation through affordable, practical technologies and inclusive economic opportunities. However, the challenge remains in scaling these innovations, because widespread adoption is critical. Although these efforts are promising, sustained commitment is essential for real change.

In summary, businesses enhance climate resilience at the BoP (bottom of the pyramid) by addressing specific vulnerabilities through tailored solutions that combine innovation, affordability, and inclusivity. By aligning profit motives with social impact, not only mitigate climate risks for BoP communities; but also unlock sustainable growth opportunities. Moreover, balancing profit with social impact is always a challenge for businesses. However, initiatives like these toward building resilience against climate change make adaptation a cornerstone of successful business strategies.

5.7 Socio-Economic Barriers to Embracing Climate Technologies and Possible Negative Externalities

The incorporation of sophisticated climate technologies in rural (and economically disadvantaged) areas faces significant socio-economic challenges. Elevated initial expenses, combined with the need for sufficient supporting infrastructure, frequently render these technologies inaccessible to low-income families and smallholder farmers; this limitation curtails their potential effectiveness (Westermann et al., 2018). Many rural communities, however, find themselves lacking access to credit resources or governmental subsidies, which are essential for funding these innovations. In addition, the knowledge gap concerning the use and upkeep of advanced systems such as renewable energy solutions, precision agriculture tools, or AI-driven resource management technologies intensifies the existing dilemma. Although socio-cultural elements contribute to the issue, resistance to new practices often arises from a lack of awareness or a perceived misalignment with traditional techniques. This complex array of challenges significantly hinders widespread adoption (Partey et al., 2018).

The integration of these technologies (however) presents potential negative externalities. For example, solar and wind energy projects, which are designed to decrease carbon footprints, can (on a large scale) result in land acquisition disputes, the displacement of communities, and ecological degradation (Stock, 2022). Similarly, the transition to electric vehicles may intensify mining for rare earth elements, causing environmental harm and exploitation of labor in resource-rich regions (Pretty & Bharucha, 2014). Although there are benefits, an over-reliance on technological solutions might overshadow the significance of traditional, community-driven adaptation strategies. This could erode local knowledge systems and create dependency on external systems. Thus, while climate technologies hold immense promise, it is crucial to address these socio-economic barriers and mitigate their unintended consequences, because doing so will ensure equitable and sustainable adoption.

5.8 Conclusion

In Conclusion, climate change is one of the greatest challenges that vulnerable communities face today, more so in developing nations, such as India where dependency on natural resources remains high. These physical, social, and economic contexts determine not only the vulnerability of people to climate change but also their capacity to reduce their exposure and risk, so subnational risk assessments and appropriate adaptation measures are essential. In addition mostly in case of climate vulnerability, the affected sectors are those that operate within the realm of climate change sensitivity like agriculturally based communities which have more vulnerable women group. The way of addressing these vulnerabilities is the need to adopt preventive measures which aim at increasing the level of community preparedness through education, gender, infrastructure, and resources. However, possible solutions for key coping mechanisms, for example, migration, changes in agriculture practices, and other coping diversification strategies all classified as sustainable require comprehensive policies to sustain them.

In parallel with these community transformations, the map of Indian business is developing dramatically. There is a rising awareness of climate change among companies and startups, and the industry increasingly is bringing sustainability to its focus. Biodegradable startups such as Buyofuel, Ace Green Recycling, and Green-Jams, and the giant UltraTech Cement, show that profitability and sustainability are not mutually exclusive. These businesses are using IoT, big data, and biofuel processing as well as ensuring they meet their environmental goals and obligations toward international accords including SBTi and GCCA Climate Ambition 2050.

Nature-based solutions (NCS), afforestation, and carbon capture are strengthening corporate sustainability efforts, but they require greater support from governments and expanded public–private partnerships due to legal and quality challenges.

At last, this book chapter provided an analysis of the collective actions of Indian businesses and communities, highlighting the growing recognition of their crucial role in managing climate change impacts and pursuing adaptation strategies. By introducing new policies that encourage innovation and collaboration and stewardship of sustainability, organizations are not only protecting business interests but also significantly contributing toward the welfare of the global community. At the same time, preparing the necessary resources, knowledge, and policies is crucial for ensuring the proper development of future generations, both in India and globally. Ultimately, when businesses, governments, and communities come together, they create a balanced and sustainable future for the economy, environment, and society.

References

Abu, M., Codjoe, S. N. A., & Sward, J. (2014). Climate change and internal migration intentions in the forest-savannah transition zone of Ghana. *Population and Environment, 35*(4), 341–364. https://doi.org/10.1007/s11111-013-0191-y

Acosta-Michlik, L., & Espaldon, V. (2008). Assessing vulnerability of selected farming communities in Phillipines based on a behavioural model of agent's adaptation to Global Environmental Change. *Global Environmental Change, 18*(4), 554–563.

Acosta-Michlik, L., Kelker, U., & Sharma, U. (2008). A critical overview: Local evidence on vulnerabilities and adaptations to global environmental chnage in developing countries. *Global Environmental Change, 18*(4), 539–542.

Adams, R., McCarl, B., Segerson, K., Rosenzweig, C., Bryant, K., Dixon, B., et al. (1999). The economic effect of climate change on US agriculture. In R. Mendelsohn & J. Neumann (Eds.), The economic impact of climate change on the economy of the United States. Cambridge University Press.

Adeniji-Oloukoi, G., Urmilla, B., & Vadi, M. (2013). Households' coping strategies for climate variability related water shortages in Oke-Ogun region, Nigeria. *Environmental Development, 5*, 23–38.

Adger, W. N. (2003). Social capital, collective action and adaptation to climate change. *Economic Geography, 79*(4), 387–404.

Adger, W. N. (2006). Vulnerability. *Global Environmental Change, 16*(3), 268–281.

Afifi T, Liwenga E, Kwezi L. (2014). Rainfall-induced crop failure, food insecurity and out-migration in same-Kilimanjaro, Tanzania. *Clim Dev 6*(1), 53–60. https://doi.org/10.1080/17565529.2013.826128

Agrawal, A. (2001). Common property institutions and sustainable governance of resources. *World Development, 29*, 1649–1672.

Aguilar, L. (2009). Women and climate change: Vulnerabilities and adaptive capacities. http://www.worldwatch.org/files/pdf/SOW09_CC_women.pdf. (Accessed 25 august 2024)

Ajani, E. N., Onwubuya, E. A., Mgbenka, R. N., (2013). Approaches to economic empowerment of rural women for climate change mitigation and adaptation: Implications for policy. *International Journal of Advance Agricultural Research, 1*, 37e42.

Alexander, K. S., Millar, J., & Lipscombe, N. (2010). Sustainable development in the uplands of Lao PDR. *Sustainable Development, 18*(1), 62–70.

Alfthan, B., Skaalvik, J., Johnsen, K., Sevaldsen, P., Nellemann, C., Verma, R., & Hislop, L. (2011). *Women at the frontline of climate change: Gender risks and hopes.*

Ali, J., & Kapoor, S. (2008). Farmers' perception on risks in fruits and vegetables production. *Agricultural Economics Research Review, 21*, 317–326.

Alston, M. (2014, November). Gender mainstreaming and climate change. In *Women's Studies International Forum, Vol. 47* (pp. 287–294). Pergamon.

Arora-Jonsson, S. (2011). Virtue and vulnerability: Discourses on women, gender and climate change. *Global Environmental Change, 21*(2), 744–751.

Behara, B., & Vaswani, R. T. (2007). Household perception of climate change in Leh, India. Vidyasagar University. *Journal of Economics, 12*(1), 82–89.

Berman, R., Quinn, C., & Paavola, J. (2013). "Identifying drivers of household coping strategies to multiple climatic hazards in Western Uganda: Implications for adapting to future climate change", Policy Working Paper No 149, Centre for Climate Change Economics.

Black, R., Adger, W. N., Arnell, N. W., Dercon, S., Geddes, A., & Thomas, D. (2011). The effect of environmental change on human migration. *Global Environmental Change, 21*(SUPPL. 1), S3–S11. https://doi.org/10.1016/j.gloenvcha.2011.10.001

Brody, S. D., Zahran, S., Vedlitz, A., & Grover, H. (2008). Examining the relationship between physical vulnerability and public perceptions of global climate change in the United States. *Environment and Behavior, 40*(1), 72–95.

Bryan, E., Deressa, T. T., Gbetibouo, G. A., & Ringler, C. (2009). Adaptation to climate change in Ethiopia and South Africa: Options and constraints. *Environmental Science Policy, 12*, 413–426.

Bryan, E., Ringler, C., Okoba, B., Roncoli, C., Silvestri, S., & Herrero, M. (2010), "Coping with climate variability and adapting to climate change in Kenya: Household and community strategies and determinants", Adaptation of Smallholder Agriculture to Climate Change in Kenya, World Bank.

Byg, A., & Salick, J. (2009). Local perspectives on a global phenomenon-climate change in eastern Tibetan villages. *Global Environmental Change, 19*(2), 156–166.

CDKN (Climate and Development Knowledge Network). (2011). Climate change and poverty reduction. Policy Brief. http://cdkn.org/wp-content/uploads/2011/08/CDKNpoverty-reduction FINAL.pdf.

Chigavazira, B. M. (2012), "Adapting to climate change to achieve household food security: A case study of small-scale farmers at Dzindi smallholder irrigation scheme in the Limpopo province of South Africa", MSc thesis, University of Fort Hare.

Christian Pirzer, Endeva, iBAN. (2015, November 30). *How Businesses can Increase Climate Resilience of BoP Communities.* iBAN. https://www.inclusivebusiness.net/ib-voices/how-businesses-can-increase-climate-resilience-bop-communities

Christopher Beck, jen@environmentalleader.com. (2024, July 9). Climate change and big business: Bridging the gap. *Environment+Energy Leader*. https://www.environmentenergyleader.com/stories/climate-change-and-big-business-bridging-the-gap,25787

Craven, L. K. (2015). Migration-affected change and vulnerability in rural Vanuatu. *Asia Pacific Viewpoint, 56*(2), 223–236. https://doi.org/10.1111/apv.12066

Curtol, F., Decarli, S., Di Nicola, A., & Savona, E. U. (2004). Victims of human trafficking in Italy: A judicial perspective. *International Review of Victimology, 11*(1), 111–141.

Davidson, O., Halsnaes, K., & Huq, S. (2003). The development and climate nexus: The case of sub-Saharan Africa. *Climate Policy, 3*, S97–S113.

Deressa, T., Hassan, R. M., Alemu, T., Yesuf, M., & Ringler, C. (2008). Analyzing the determinants of farmers' choice of adaptation methods and perceptions of climate change in the Nile Basin of Ethiopia. IFPRI Discussion Paper 00798.

Easterling, W. E., Crosson, P. R., Rosenberg, N. J., McKenney, M. S., Katz, L. A., & Lemon, K. M. (1993). Agricultural impacts of and responses to climate change in the Missouri-Iowa-Nebraska region. *Climatic Change, 24*(1–2), 23–61.

Ebi, K. L., Woodruff, R., Von Hildebrand, A., & Corvalan, C. (2007). Climate change related health impacts in the Hindu Kush—Himalayas. *EcoHealth, 4*, 264–270.

Ellis, F. (1999). *Rural livelihood diversity in developing countries: Evidence and policy implications.* Natural Resource Perspectives 40(8).

Escarcha, J. F., Lassa, J. A., & Zander, K. K. (2018). Livestock under climate change: A systematic review of impacts and adaptation. *Climate, 6*(3), 54.

Etzold B., Ahmed A. U., Hassan S. R., Neelormi, S. (2014). Clouds gather in the sky, but no rain falls. Vulnerability to rainfall variability and food insecurity in Northern Bangladesh and its effects on migration. *Clim Dev, 6*(1), 18–27. https://doi.org/10.1080/17565529.2013.833078

FAO. (2018). Safe access to fuel and energy (SAFE) strengthening resilience of 2 crisis-affected populations. https://www.fao.org/3/i8012en/I8012EN.pdf.

Feng, S., Krueger, A. B., & Oppenheimer, M. (2010). Linkages among climate change, crop yields and Mexico-US cross-border migration. *Proceedings of the National Academy of Sciences, 107*(32), 14257–14262. https://doi.org/10.1073/pnas.1002632107

Gallopin, G. C. (2006). Linkages between vulnerability, resilience, and adaptive capacity. *Global Environmental Change, 16*(3), 235–316.

Grown, C., Gupta, G. R., Kes, A. (2006). Taking action to empower women: UN millennium project report on education and gender equality. *Global Urban Development, 2*(1), 1e19.

Haigh, C., Vallely, B. (2010). Gender and the climate change agenda: The impacts of climate change on women and public policy. Women's Environmental Network © 2010 20 Club Row.

Henry, B. K., Eckard, R. J., & Beauchemin, K. A. (2018). Adaptation of ruminant livestock production systems to climate changes. *Animal, 12*(s2), s445–s456.

Herrero, M., Henderson, B., Havlík, P., Thornton, P. K., Conant, R. T., Smith, P., & Stehfest, E. (2016). Greenhouse gas mitigation potentials in the livestock sector. *Nature Climate Change, 6*(5), 452–461.

Herrero, M., Thornton, P. K., Mason-D'Croz, D., Palmer, J., Benton, T. G., Bodirsky, B. L., & West, P. C. (2020). Innovation can accelerate the transition towards a sustainable food system. *Nature Food, 1*(5), 266–272.

Hill, C. (2011). Enabling rural Women's economic empowerment: Institutions, opportunities, and participation. In Background Paper for Expert Group Meeting, Accra, 20e23 September. http://www.un.org/womenwatch/daw/csw/csw56/egm/Hill-BP-1-EGM-RW-Sep-2011_CH.pdf.

Hulme, D., & Shepherd, A. (2003). Conceptualizing chronic poverty. *World Development, 31*(3), 403–423.

Intergovernmental Panel on Climate Change (IPCC). (2001). *Climate change 2001: The scientific basis.* Cambridge University Press.

Intergovernmental Panel on Climate Change (IPCC). (2007a). *Climate change 2007: Synthesis report. Contribution of Working Groups I, II and III to the Fourth Assessment Report of the Intergovernmental Panel on Climate Change* (Core Writing Team, R. K. Pachauri & A. Reisinger, Eds.). IPCC. https://www.ipcc.ch/report/ar4/syr/

Intergovernmental Panel on Climate Change (IPCC). (2007b). *Climate Change 2007: Impacts, adaptation and vulnerability. Contribution of Working Group II to the Fourth Assessment Report of the Intergovernmental Panel on Climate Change* (M. L. Parry, O. F. Canziani, J. P. Palutikof, P. J. van der Linden, & C. E. Hanson, Eds.). Cambridge University Press. https://www.ipcc.ch/report/ar4/wg2

Jones Christensen, L., Siemsen, E., & Balasubramanian, S. (2015). Consumer behavior change at the base of the pyramid: Bridging the gap between for-profit and social responsibility strategies. *Strategic Management Journal, 36*(2), 307–317.

Karl, T. R., Melillo, J. M., & Peterson, T. C. (2009). *Global climate change impacts in the United States.* Cambridge University Press.

Keskitalo, E. C. H., & Kulyasova, A. A. (2009). The role of governance in community adaptation to climate change. *Polar Research, 28*(1), 60–70.

Krishna Chaitanya, S. V. (2023, December 5). 'Michaung is clear signature of climate change.' *The New Indian Express*. https://www.newindianexpress.com/nation/2023/Dec/05/michaung-is-clear-signature-of-climate-change-2638741.html

Kumar, A., & Mohanasundari, T. (2024). Assessing district-level climate vulnerability in Madhya Pradesh, Central India: An integrated environmental and socio-economic approach. *Theoretical and Applied Climatology, 155*(4), 3449–3471.

Kumar, K. K., Rajagopalan, B., & Cane, M. A. (1999). On the weakening relationship between Indian monsoon and ENSO. *Science, 284*(5423), 2156–2159.

Kvanneid, A. J. (2021). Perceptions of climate change from north India: An ethnographic accoun. Routledge. https://www.routledge.com/Perceptions-of-Climate-Change-from-North-India-An-EthnographicAccount/Kvanneid/p/book/9780367421434

Lambrou, Y., & Piana, G. (2006). *Gender: The missing component of the response to climate change* (pp. 1–58). FAO.

Lambrou, Y., & Nelson, S. (2010). *Farmers in a changing climate: Does gender matter?*

Legide, Y. Y., Feyissa, G. S., & Karo, T. M. (2024). Revitalizing indigenous practices employed by farmers to reduce agriculture's vulnerability to climate change: A systematic review. *Journal of Environmental Studies and Sciences, 14*(2), 400–414.

Locatelli, B., Brockhaus, M., Buck, A., Thompson, I., Bahamondez, C., Murdock, T., Roberts, G., & Webbe, J. A. (2010). "Forests and society: Responding to global drivers of change". In M. Gerardo, K. Pia, G. Glenn, A. René, K. Markku, & V. Jari (Eds.), *IUFRO World Series* (25). IUFRO.

Lutz, W., Muttarak, R., Striessnig, E. (2014). Universal education is key to enhanced climate adaptation. *Science, 346*(6213). https://doi.org/10.1126/science.1257975.

Mendelsohn, R., & Dinar, A. (2003). Climate, water, and agriculture. *Land Economics, 79*(3), 328–341.

Mendelsohn, R., Nordhaus, W., & Shaw, D. (1994). The impact of global warming on agriculture: A Ricardian analysis. *American Economic Review, 84*(4), 753–771.

Millennium Ecosystem Assessment. (2005). *Ecosystems and human well-being: Synthesis*. Island Press.

Nanda, N. (2009). *Climate change and trade policy: A preliminary discussion*. Asia Pacific Research and Training Network on Trade (UN ESCAP). Issue no. 4.

Nath, P. K., & Behera, B. (2011). A critical review of impact of and adaptation to climate change in developed and developing economies. *Environment, Development and Sustainability, 13*, 141–162.

Njenga, M., Iiyama, M., Terada, Y., Kitenge, D., Gitau, J. K., Kinuthia, R., & Mendum, R. (2024). "The problem is a lack of firewood": Charcoal briquettes for cooking energy in refugee and host communities. *Social Sciences & Humanities Open, 9*, Article 100852.

Nuorteva, P., Keskinen, M., & Varis, O. (2010). Water, livelihoods and climate change adaptation in the Tonle Sap Lake area, Cambodia: Learning from the past to understand the future. *Journal of Water and Climate Change, 1*(1), 87–101.

Nwoke, E. A., Ibe, S. N. O. (2014). Climate change impact on the health of African women and adaptation strategies. *International Journal of Global Warming Climate Change, 1*(1).

O'Brien, K., Eriksen, S., Schjolden, A., & Nygaard, L. P. (2007). What's in a word? Conflicting interpretations of vulnerability in climate change research. *Climate Policy, 7*, 73–88.

Painter, M. A., Shah, S. H., Damestoit, G. C., Khalid, F., Prudencio, W., Chisty, M. A., ... & Wilhelmi, O. (2024). A systematic scoping review of the social vulnerability index as applied to natural hazards. *Natural Hazards*, 1–92.

Partey, S. T., Zougmoré, R. B., Ouédraogo, M., & Campbell, B. M. (2018). Developing climate-smart agriculture to face climate variability in West Africa: Challenges and lessons learnt. *Journal of Cleaner Production, 187*, 285–295.

Pramova, E., Locatelli, B., Djoudi, H., & Somorin, O. A. (2012). Forests and trees for social adaptation to climate variability and change. *Wires Climate Change, 3*, 581–596.

Press Information Bureau, Government of India. (2022, February 2). India has taken lead to raise the issue of climate finance at COP26. *Calls for New and Additional Finance, Grants Over Loans, Balanced between Mitigation and Adaptation.* https://www.pib.gov.in/PressReleasePage.aspx?PRID=1845822

Pretty, J., & Bharucha, Z. P. (2014). Sustainable intensification in agricultural systems. *Annals of Botany, 114*(8), 1571–1596.

Prowse, M., Grist, Natasha, Sourang, Cheikh (2009). Closing the gap between climate adaptation and poverty reduction frameworks. Overseas Development Institute Project Briefing No. 21. http://www.odi.org/sites/odi.org.uk/files/odi-assets/publications-opinion-files/4297.pdf.

Quinn, C. H., Ziervogel, G., Taylor, A., Takama, T., & Thomalla, F. (2011). Coping with multiple stresses in rural South Africa. *Ecology and Society, 16*(3), 1–20.

Rodgers, Y., Menon, N. (2013). A meta-analysis of land rights and women's economic well-being the roadmap for action—a tool for funders, practitioners and policymakers to promote women's economic empowerment. http://www.womeneconroadmap.org/sites/default/files/Rodgers%20and%20Menon_Survey%20of%20Land%20Rights%20and%20Women.pdf

Sakdapolrak, P., Promburom, P., & Reif, A. (2014). Why successful in situadaptation with environmental stress does not prevent people from migrating? Empirical evidence from Northern Thailand. *Climate and Development, 6*(1), 38–45. https://doi.org/10.1080/17565529.2013.826129

Saldana-Zorrilla, S. O. (2008). Stakeholders' views of reducing rural vulnerability to natural disasters in Southern Mexico: Hazard exposure and coping and adaptive capacity. *Global Environmental Change, 18*(4), 583–597.

Salman, R. A., Ferdinand, T., Carter, R., & Choularton, R. (2019). Transformative adaptation in livestock production systems.

Sandhya, M., & Dashora, P. K. (2003, January). Communication strategies for technological empowerment of farm women in post-harvest management. In *International Conference on Communication for Development in the Information Age. Extending the Benefits of Technology for All* (pp. 7–9).

Schipper, E., Lisa, F. (2007). *Climate change adaptation and development: Exploring the linkages.* Tyndall Centre Working Paper 107.Tyndall Centre for Climate Change Research, University of East Anglia and South East Asia START Regional Centre, Norwich, UK and Bangkok, Thailand.

Sejian, V., Gaughan, J., Baumgard, L., Prasad, C. S. (Eds.). (2015). *Climate change adaptation livestock: Impact on and mitigation.* Springer. https://doi.org/10.1007/978-81-322-2265-1

Sen, A. (1999). *Development as Freedom.* Oxford University Press.

Shackleton, S. E., Campbell, B., Lotz-Sisitka, H., & Shackleton, C. M. (2008). Links between the local trade in natural products, livelihoods and poverty alleviation in a Semi-Arid region of South Africa. *World Development, 36*(3), 505–526.

Simms, A., Magrath, J., Reid, H. (2004). Up in smoke? International institute for environment and development new economics foundation (38 pp).

Singh, S. (2020). Farmers' perception of climate change and adaptation decisions: A micro-level evidence from Bundelkhand Region, India. *Ecological Indicators, 116*, 106475.

Smith, B., & Diedrich, A. (2024). A systematic review of current progress in community based vulnerability assessments. *Regional Environmental Change, 24*(1), 21.

Smith, K. (2000). *Environmental hazards: Assessing risk and reducing disaster*. Routledge.

Sohngen, B., Mendelsohn, R. (1998). Valuing the impact of large-scale ecological change in a market: The effect of climate change on U.S. timber. *The American Economic Review, 88*(4), 686–714.

Solar, W. R. (2010). Rural women, gender, and climate change. A literature review and invited perspectives on climate change impacts and processes of adaptation in Cambodia. http://www.climateadapt.asia/upload/publications/files/4E4E347325B40LITERATURE_REVIEW_-_Rural_Women_Gender_and_Climate_Change.pdf. (Accessed 25 august 2024).

Stern, N. (2006). *Stern review: The economics of climate change.* Cambridge University Press.

Stock, R. (2022). Triggering resistance: Contesting the injustices of solar park development in India. *Energy Research & Social Science, 86*, Article 102464.

Stone, K., Blinn, N., & Spencer, R. (2022). Mental health impacts of climate change on women: A scoping review. *Current Environmental Health Reports, 9*(2), 228–243.

Weindl, I., Lotze-Campen, H., Popp, A., Müller, C., Havlík, P., Herrero, M., ... & Rolinski, S. (2015). Livestock in a changing climate: Production system transitions as an adaptation strategy for agriculture. *Environmental Research Letters, 10*(9), 094021.

Westermann, O., Förch, W., Thornton, P., Körner, J., Cramer, L., & Campbell, B. (2018). Scaling up agricultural interventions: Case studies of climate-smart agriculture *Agricultural Systems, 165*, 283–293.

WOCAN. (2014). Women's empowerment at the frontline of adaptation emerging issues, adaptive practices, and priorities in Nepal. http://www.wocan.org/system/tdf/Women_Empowerment_Adaptation_ICIMOD_CICERO_WOCAN%202014.pdf?file¼1&type¼node&id¼1166

World Bank Group. (2023, March 31). *India: Climate change impacts*. World Bank. https://www.worldbank.org/en/news/feature/2013/06/19/india-climate-change-impacts

World Bank Report. (WBR). (2009). South Asia: Climate change strategy.

World Bank. (2004). *The impact of energy on women's lives in rural India*. The World Bank, Joint UNDP/World Bank Energy Sector Management Assistance Programme (ESMAP).

Yadav, R. P., Singh, P., Arya, S. L., Bhatt, V. K., & Sharma, P. (2008). *Detailed Project Report (xxx) Watershed Panchkula [For Implementation under NWDPRA Scheme, Ministry of Agriculture, Govt. Of India, New Delhi]*. Central Soil & Water Conservation Research and Training Institute, Research Centre.

Falaq Mubashir is an Assistant Professor (Guest) at the National Institute of Technology, Srinagar, where she combines her academic rigor and research insight to advance the field of business management. She earned her PhD in Marketing in 2023, specializing in the application of advanced quantitative techniques to address complex marketing challenges. Her research primarily focuses on consumer behavior, data analytics, and digital marketing strategies. She is passionate about understanding how digital transformation is reshaping consumer trends and influencing marketing strategies across industries. Through her research, Dr. Mubashir aims to provide actionable insights into the nuances of marketing dynamics in the digital era, empowering businesses to adapt and thrive in a highly competitive marketplace.

Tahir A. Wani is Assistant Professor in the Department of Humanities, Social Sciences and Management, NIT Srinagar, India, wherein he teaches various management-related subjects to management and engineering students. Before joining NIT Srinagar, he was working at Business School, University of Kashmir, India. He has a rich research background and has written extensively on various management issues. He has also conducted workshops on quantitative analysis and participated in numerous workshops and conferences in reputed institutions across India.

Ramendra Singh is Professor of Marketing at IIM Calcutta. He completed PhD from IIM Ahmedabad, MBA from XLRI Jamshedpur, and B. Tech from IIT-BHU. His research has been published in Journal of Business Research, International Marketing Review, Marketing Theory, Industrial Marketing Management, Journal of Personal Selling & Sales Management, among others. He has previously worked in sales and marketing responsibilities across different industries for several years. He has authored a textbook on Sales and Distribution Management which has been adopted in MBA programs at various IIMs and other Business Schools. He has completed several consultancy projects, and conducted several training programs for senior and middle management cadre at several organizations.

Chapter 6
Sustainability in Subsistence Marketplaces and Beyond: Addressing Challenges Through the Bottom-Up Approach

Madhu Viswanathan

Abstract Subsistence marketplaces are likely to be most affected by environmental issues. Our work is grounded in more than two and a half decades of work with low-literate, low-income consumers, entrepreneurs, and marketplaces—i.e., subsistence marketplaces. This chapter provides a comprehensive overview of key insights that have emerged at the intersection of sustainability and subsistence. Unique to our approach is being bottom-up, beginning at the micro-level behavioral understanding of subsistence marketplaces as they intersect with the environment, consumption, and conservation. This is in contrast to the customary vantage point in sustainability of being top-down. We bridge research, education, and practice, dubbing our approach symbiotic academic-social enterprise, distinct from action research. The interplay between subsistence and sustainability using a bottom-up vantage point leads to important new insights. The chapter is organized around two major sections—understanding subsistence marketplaces and sustainability, and addressing challenges through product design and development, business models, education, academic-social enterprise, and sustainable development goals.

Keywords Subsistence marketplaces · Bottom-up approach · Marketplace literacy · Sustainability literacy · Sustainability

This chapter explores how the stream of work on subsistence marketplaces with its unique bottom-up approach and micro-behavioral insights can address sustainability

---

M. Viswanathan (✉)
Department of Marketing, College of Business Administration, Loyola Marymount University, Los Angeles, CA, USA
e-mail: marketplaceliteracy@gmail.com

Department of Business Administration, Gies College of Business, University of Illinois, Urbana-Champaign, Champaign, IL, USA

Nottingham University Business School, Nottingham, UK

R. Singh and T. A. Wani (eds.), *Marketing to the Base of the Pyramid*,
https://doi.org/10.1007/978-981-96-7105-2_6

challenges. Climate change disproportionately affects those with the least (Intergovernmental Panel on Climate Change [IPCC], 2014), impacting agricultural and food systems (Brown & Funk, 2008). Those living in low-income contexts are likely to be affected the most (Barnett, 1997), increasing disease, malnutrition, and decreasing employment and food supply for subsistence communities. This chapter provides a comprehensive overview of key insights from the stream of work on subsistence marketplaces (Viswanathan, 2013; Viswanathan & Rosa, 2007). It is grounded in more than two and a half decades of work with low-literate, low-income consumers, entrepreneurs, and marketplaces—i.e., subsistence marketplaces. It covers important challenges at the intersection of sustainability and subsistence as well as possible approaches to addressing them.

Unique to our approach is bottom-up, micro-level behavioral understanding of subsistence marketplaces as they intersect with the environment, consumption, and conservation. Keeping with this stream, we adopt a bottom-up vantage point. Although this intersection has been studied at macro and micro household levels, such as the use of natural resources in rural settings we aim for in-depth insight into psychological and socio-cultural facets. In this regard, both global and local perspectives on the effects of climate change and solutions toward more sustainable living must be considered (UNESCO, 2017). We argue that the customary vantage point in sustainability is top-down, such as from the perspectives of government, non-profit organizations, businesses, and development organizations.

Our approach bridges research, education, and practice. It has been entitled symbiotic academic-social enterprise, distinct from action research, i.e., characterized by sustained immersion in communities and a mutually enriching relationship between research and practice (Viswanathan et al., 2020). This is of particular importance in the arena of sustainability, given the myriad actors from different sectors needed to create change. Deprivation being in multiple realms, the complex interrelationships between domains of subsistence are central to understanding sustainability. The interplay between subsistence and sustainability using a bottom-up vantage point leads to important new insights. The chapter is organized into major sections on understanding subsistence marketplaces and sustainability, and addressing challenges through product design and development, business models, education, and academic-social enterprise, to achieve sustainable development goals.

6.1 Understanding Sustainability and Subsistence Marketplaces

6.1.1 Subsistence Marketplaces

The subsistence marketplace approach evolved over more than two decades—"subsistence" referring to the broad range of low-income contexts, from extreme poverty to the edge of lower-middle income at the other end (Viswanathan, 2013;

Viswanathan & Rosa, 2007; Fig. 6.1). The term, "marketplaces," reiterates the need to understand pre-existing marketplaces bottom-up, inside-out and for its own sake, distinct from the term, "markets," which can trigger a top-down, outside-in, means to other ends (e.g., new markets) mindset. This inside-out approach is particularly germane and complementary to the customary top-down approaches to sustainability, typically construed at broad, global level and from the vantage of large organizations and nations.

Subsistence marketplaces—consumers, entrepreneurs, and communities in a broad range of low-income—are characterized by the day-to-day uncertainties they face without much margin for error (Fig. 6.2; Viswanathan, 2013). Living close to the environment, they are particularly susceptible to climate change which can reverse progress (Ahmed et al., 2009). The effects of this global phenomenon vary by geographic regions, and different subsistence contexts, as communities live with uncertainties, being vulnerable to natural disasters, drought, uninhabitable heat, and

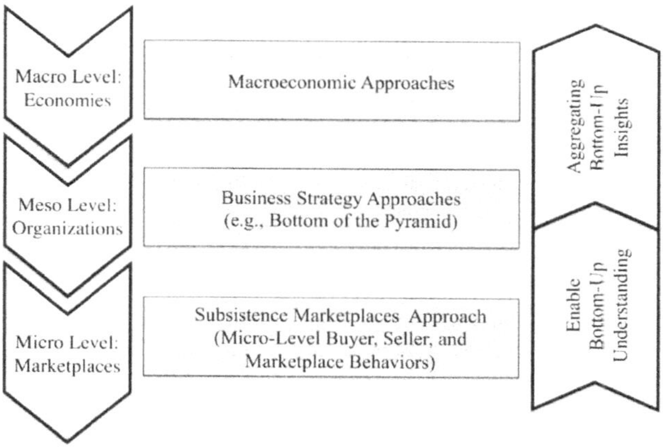

Fig. 6.1a The subsistence marketplaces stream—Perspectives on poverty and business. Adapted from Viswanathan and Venugopal (2015)

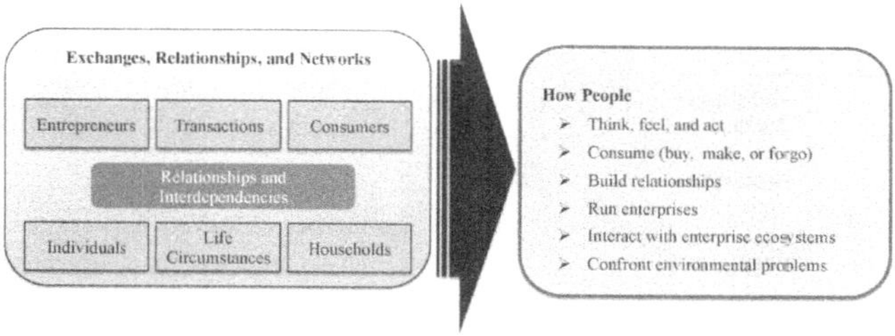

Fig. 6.1b The subsistence marketplaces stream—Microlevel phenomena. Adapted from Viswanathan and Venugopal (2015)

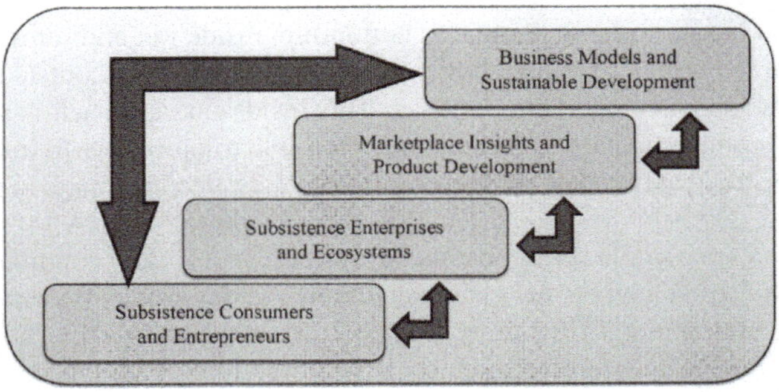

Fig. 6.1c The subsistence marketplaces stream—Bottom-up beginning at micro-level. Adapted from Viswanathan and Venugopal (2015)

deforestation, accelerated by climate change. Domains of subsistence, such as food and water, may be severely affected, forcing large-scale migrations (Meze-Hausken, 2000).

Along with uncertainties inherent in day-to-day life overlaid with environmental uncertainties, another source of uncertainty is the timely availability of sustainable solutions. These contexts are often the most difficult to reach with potential solutions. Hindering such reach are obstacles due to lack of access and infrastructure. Also relevant here is the confluence of unfamiliarities among those in relatively resource-rich settings, such as practitioners, researchers, and educators (Viswanathan, 2013; Fig. 6.2). Subsistence marketplaces are also characterized by distinctiveness of contexts, with myriad differences. Such differences may stem from socio-cultural

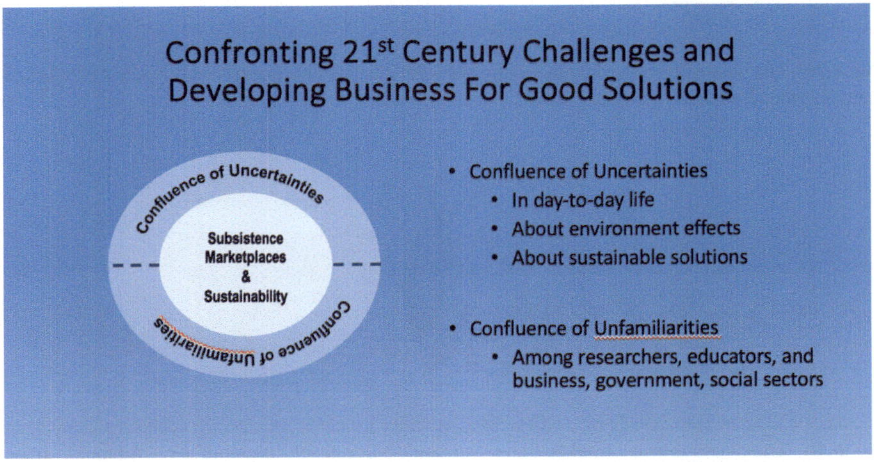

Fig. 6.2a Rationale for the bottom-up approach to subsistence marketplaces

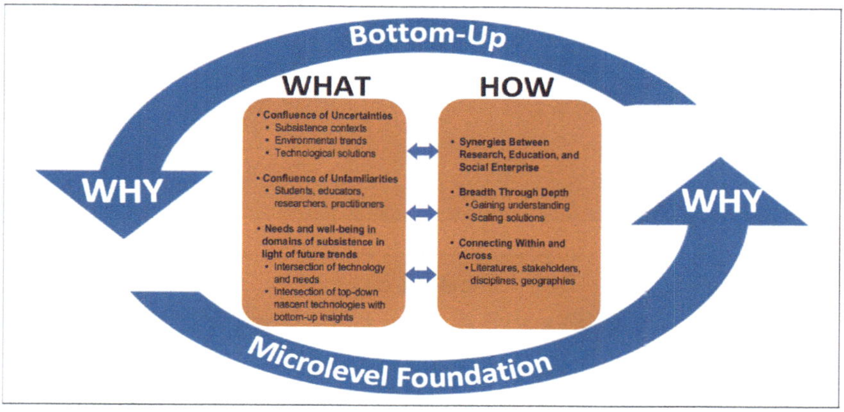

Fig. 6.2b Pathways forward for the subsistence marketplaces stream

factors or due to lack of access to information, education, and infrastructure that levels the playing field to some degree for relatively resource-rich contexts. They arise from local culture, political stability, social hierarchies, and other consequences of extreme resource constraints. Thus, inherent uncertainty and distinctiveness characterize these settings on the inside, along with unfamiliarity from the outside. Compounding the uncertainty and distinctiveness of subsistence marketplaces, and unfamiliarity about them are the varying effects of climate change (Viswanathan et al., 2020). Indeed, sustainability even in relatively resource-rich settings can be argued to be characterized as being at a similar confluence of uncertainties and a confluence of unfamiliarities, with distinctiveness in environmental effects playing out across contexts (Fig. 6.2).

This stream has brought out how individuals in these contexts are not only materially poor but face cognitive and emotional constraints due to low literacy yet can be socially or relationally rich (Viswanathan et al, 2012; Fig. 6.3). Due to low income and the closely associated foundational factor of low literacy, individuals have difficulty with abstractions and engage in concrete thinking (Viswanathan et al., 2005; Fig. 6.3). Such thinking manifests in difficulties in understanding concepts such as sustainability (e.g., climate change or global warming), and in abstracting across time, space, and social circles (e.g., envisioning global environmental phenomena into the future). It also manifests in difficulties in combining pieces of information to make more abstract inferences, such as trade-offs or gauging value (plastic or firewood use and impact on future health and social sustainability). Furthermore, making causal judgments involving concepts at varying levels of abstraction are very difficult (actions today and impact on the environment in the future; or climate change over time due to past actions). Therefore, sequential causation and tipping points in the arena of sustainability are difficult to grasp (e.g., impact of climate on species extinction and further downstream consequences). Thus, spatial social, temporal, and also hypothetical distances (Trope & Liberman, 2010) are difficult to envision

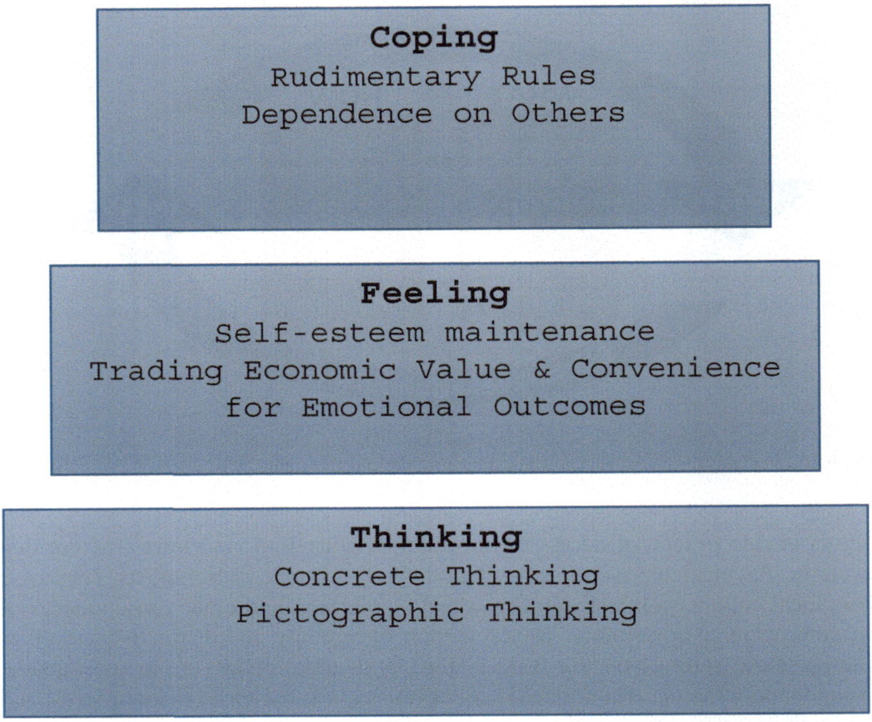

Fig. 6.3a Thinking, feeling, and coping in subsistence marketplaces. Adapted from Viswanathan et al. (2005)

and comprehend. In this sense, sustainability issues at societal and global levels are among the most abstract in being distant on all these dimensions—at a far distance spatially, socially, temporally, and hypothetically in imaging future scenarios.

These effects play out over long periods of time and distant spaces, with subtle changes that are not easily relatable from personal experiences. Juxtaposed with the very constraints we have uncovered in our work, the difficulties in grasping issues relating to sustainability, whereas true for larger population, are accentuated for subsistence marketplaces.

Individuals also fall back on the visual sensory mode through pictographic thinking (Viswanathan et al., 2005, 2009a, 2009b; Fig. 6.3), picturing brand names, visualizing amounts to buy, or manipulating mental images of currency bills to add or subtract. Visualizing or picturing is sometimes particularly difficult in the arena of sustainability, such as when pertaining to frequencies of extreme weather events over time and their impact on agriculture; or subtle differences due to climate change that occur over a period of time.

Central for subistence consumers in what are mundane shopping activities for others, is self-esteem and self-confidence, fearing exposure and even humiliation

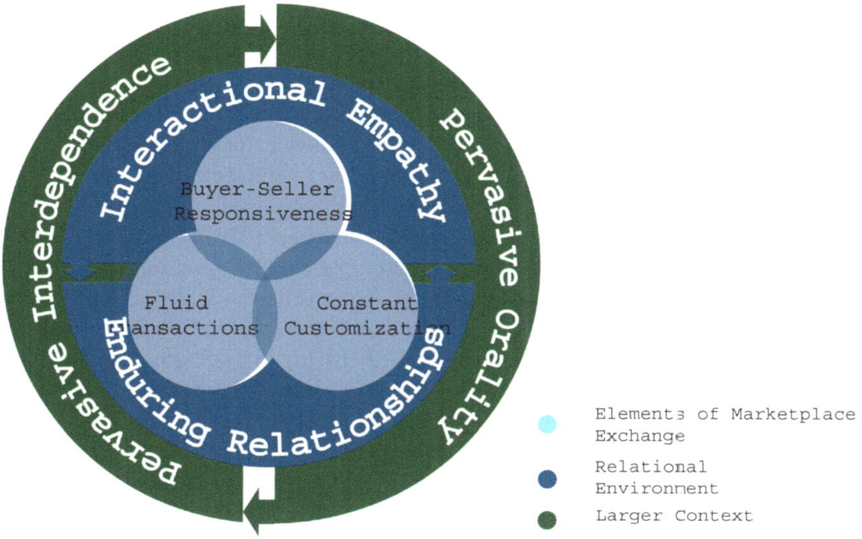

Fig. 6.3b Marketplaces in subsistence. Adapted from Viswanathan et al. (2012)

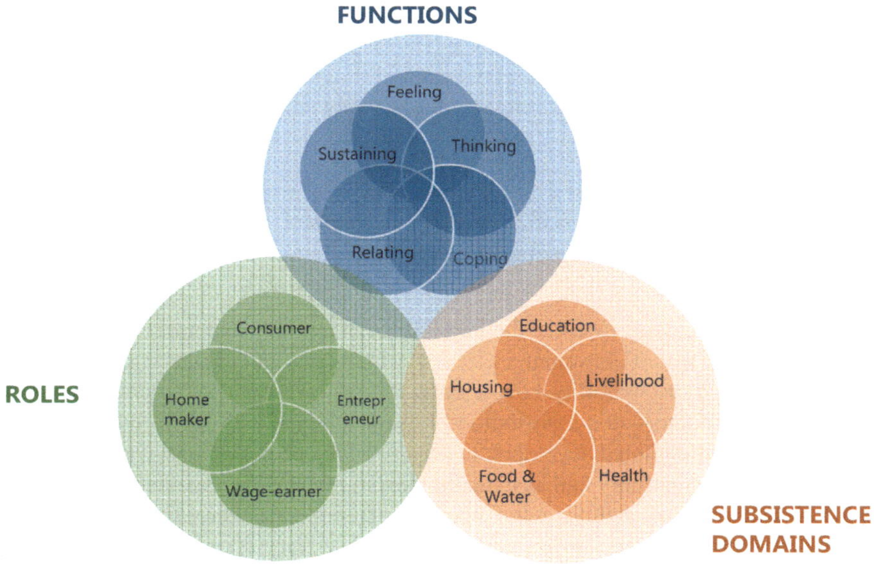

Fig. 6.3c Intertwined vulnerabilities in subsistence marketplaces. Adapted from Viswanathan et al. 2024a)

(Viswanathan et al., 2005; Fig. 6.3). Thus, marketplace interactions carry added anxieties and fears, which further inhibit exposure, learning, and development of skills and knowledge. This again can be accentuated in nascent arenas such as sustainability fraught with uncertainties. Affective and cognitive constraints can lead to a variety of coping behaviors such as avoidance, dependence, and rudimentary heuristics. Yet, the potential for relational or social richness is noteworthy. These contexts have been characterized as intensely personal, with one-on-one interactions, where fluid, and customized exchanges unfold (Viswanathan et al., 2012; Fig. 6.3). Marketplace relationships may be enduring and even empathetic, playing out in a context of interdependence and oral communications. Noteworthy is duality of consumer-entrepreneur as the same individual is buyer and seller, referred to as the term "subsistence consumer merchants" (Viswanathan et al., 2010; Fig. 6.3). Sharing and learning from common adversities, individuals may display empathy for both roles. Additionally, the bottom-up approach has been central in understanding what individuals "sustain" bottom-up, as discussed in the next section.

A number of relevant issues are covered in the section on addressing sustainability and mentioned here for completeness and addressed subsequently. Through the stream of subsistence marketplaces, a bottom-up product development process has evolved (Viswanathan, 2016; Viswanathan et al., 2012), encompassing immersion, emersion, design, innovation, and enterprise. This stream has also led to the elucidation of bottom-up-marketing (Viswanathan et al., 2024b). Furthermore, due to blurring of products and life circumstances, the social and economic, and the market and social milieu, a sustainable market orientation wherein doing good is essential for doing has been delineated for organizations (Viswanathan et al., 2009a, 2009b).

6.1.2 Understanding Sustainability Bottom-Up in Subsistence Marketplaces

First and foremost, we emphasize the importance of a bottom-up approach to understanding sustainability (Viswanathan, 2013). This is particularly germane as issues of environmental sustainability have emerged top-down through scientific studies of human-made factors that contribute to climate change and other environmental issues. Furthermore, global development organizations naturally start with top-down definitions of sustainability and sustainable development goals. Critically important here is to understand what those at the other end of the spectrum are striving to sustain. Whereas technology is top-down, and scientific discovery about climate change is top-down, culture is bottom-up and marketplaces are bottom-up.

Unique to sustainability and environmental issues in subsistence marketplaces is how people live day-to-day close to the environment and depend on natural resources in an immediate sense (firewood, water, food, materials for shelter) (Viswanathan et al., 2014). Plastic disposal is not in a remote location but in the community or nearby; thus, the consequences for blockage of sewage or soil degradation are felt immediately. Individuals have to trade-off firewood to cook in the immediate timeframe with loss of trees and deforestation medium timeframe. Using firewood for cooking in smoke-filled dwellings leads to a variety of lung ailments. A bottom-up approach brings into focus the regional and local effects of sustainability.

On the one hand, sustainability is concrete, local, and immediate, thus, trade-offs are learned through harsh day-to-day reality. On the other, tradeoffs as well as global issues of sustainability may be difficult to grasp due to difficulties with abstractions and the constraints in thinking discussed earlier. From a bottom-up vantage point, sustainability relates to *surviving* physically, *relating* to people and to the environment, and *growing* for one's household or for children and the next generation (Fig. 6.4; Viswanathan et al., 2014). Issues relating to the environment have been categorized in terms of distance as being immediate (household), near (local), moderately distant (societal), and far (global), with people having minimal control over anything but the immediate and to some degree the near or local (Fig. 6.4). Climate change looms at the farthest distance, due to very distal human-made causes, such as globalization.

Individuals in subsistence marketplaces have to act in the immediate timeframe to meet basic necessities and lack of control over farther distances (Viswanathan et al., 2014). Nevertheless, they have a deep, aspirational drive for the next generation to grow and thrive, even if their cognitive constraints make it difficult to understand the path toward sustainability. Thus, they make constant sacrifices to enable

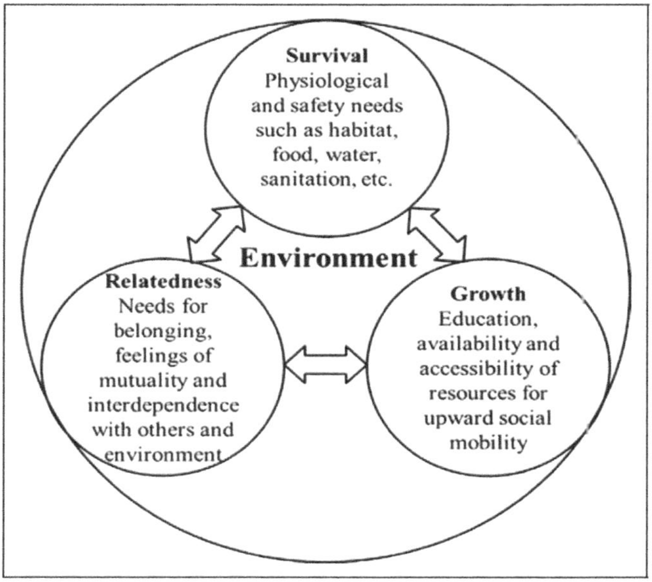

Fig. 6.4a Understanding sustainability bottom-up—Bottom-up perspective on sustainability. Adapted from Viswanathan et al. (2014)

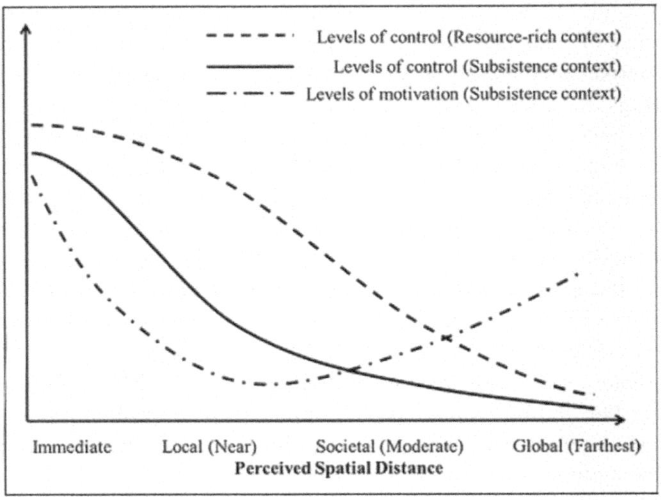

Fig. 6.4b Understanding sustainability bottom-up—Levels of control and perceived spatial distances. Adapted from Viswanathan et al. (2014)

a better future. Issues of environmental sustainability that they live through are of much concern in concrete ways when considering the future of the next generation. Despite extreme vulnerability, the immediacy of consequences does not allow for inertia. At least in the immediate spatial and social distances, the impetus for action among individuals and communities may provide a basis to build on sustainable initiatives. This is in contrast to relatively resource-rich settings where awareness about environmental issues at moderate and far distances without the immediacy that arises from living with the environment may lead to lack of action.

To summarize, subsistence communities live their environment, making impossible tradeoffs despite constraints in thinking in the abstract realm of global sustainability. They sustain through surviving, relating, and growing (Viswanathan et al., 2014). Sustainable solutions devised top-down have to meet the needs bottom-up. This assertion is amplified or accentuated when considered the rarer consumption events for those at the bottom and the strong role of culture in *sustaining* communities. The impetus to act and take concrete actions may be strong due to aspirations for the next generation and the immediate consequences of environmental problems.

6.2 Addressing Sustainability Challenges in Subsistence Marketplaces

In this next section, we discuss how sustainability challenges can be addressed through our approach. We cover a number of ways in which challenges can be addressed that have evolved in the stream of subsistence marketplaces and reflect the bottom-up approach. At a product level, bottom-up design focuses on designing sustainable goods and services with a bottom-up vantage and broad scope that goes well beyond an immediate and singular focus on the customer and need, ideal for the arena of sustainability. In terms of marketing, bottom-up marketing focuses on the *what*, *how*, and *why* of a business as well as the mindsets. In terms of business or enterprise models, the sustainable market orientation emphasizes how doing good may be essential to doing well. Marketplace literacy focuses on the critical role of education on sustainability for subsistence communities in light of the complex topics involved. Finally, symbiotic academic social enterprise represents a unique approach to academic research and practice ideally suited for the arena of sustainability, differing from even action research in emphasizing sustained immersion and coevolving practice and theory.

6.2.1 Bottom-Up Design

The subsistence marketplaces stream has evolved into a bottom-up process for sustainable product development (Viswanathan, 2016; Viswanathan & Sridharan, 2012). Iterative phases include bottom-up immersion (in ground reality), emersion (placing new insights in light of what is previously known), design, innovation, and enterprise. Bottom-up design can be distinguished from human-centric design or empathic design or even design thinking in terms of two facets—adopting a bottom-up first approach vantage point to understanding needs, designing products, and developing business plans; and having broad scope in terms of viewing not just customers but communities and the larger context. We note here that approaches such as design thinking, human-centric design, and empathic design are relatively top-down and adopt a consumer-centric approach. Through broad scope, we capture the community and the larger context as well in our approach. Thus, issues of social and environmental sustainability are explicitly considered even before the need is unpacked and articulated.

The bottom-up approach to sustainable solutions in subsistence marketplaces addresses the confluence of uncertainties, the confluence of unfamiliarities, and distinctiveness. It begins by understanding the need, as well as its drivers and the larger context elements. In contrast to identifying a need, the approach here is to understand what drives the need and also the elements of the larger setting where the need and its drivers play out, particularly relevant in addressing sustainability issues.

Uncovering aspirations, what may be seeming mundane needs in relatively resource-rich settings (e.g., good nutrition and good education for children), is another facet here.

Through such exercises as bottom-up need deconstruction, the need is unpacked and understood in-depth, in terms of what people sustain, their values and aspirations, and so forth. Furthermore, the focus is on not just on the customer but on the community as well as the larger context (Viswanathan, 2016; Fig. 6.5). Such scope allows for addressing issues of sustainability. An exclusive focus on the consumer need simply does not suffice here. This is for two primary reasons: the unique context of subsistence marketplaces blurs traditional boundaries between the individual, the household and the community, or between markets and the social milieu (Viswanathan et al., 2014); and sustainable solutions need to be broadly construed well beyond the consumer and need to address the social and environmental in addition to the economic. We developed a bottom-up process for sustainable product development through our work with 50+ yearlong projects involving large companies, startups and social enterprises (Viswanathan, 2016; Viswanathan & Sridharan, 2012) encompasses virtual and actual immersion, emersion, design, innovation, and enterprise.

Immersion, individuals move from sympathy to informed empathy (Viswanathan. Needs are articulated in unique rather than preconceived ways through our approach. Bottom-up need deconstruction involves taking the perspective of the consumer in a most granular way, an exercise where needs are categorized into sub-needs and beyond (Fig. 6.5). Again, the ideal is to understand needs in terms of their components as well as in terms of what causes them. Another exercise involves top-down need deconstruction, thus taking the perspective of the organization. Rather than claim the superiority of bottom-up, we aim to interweave it with the top-down.

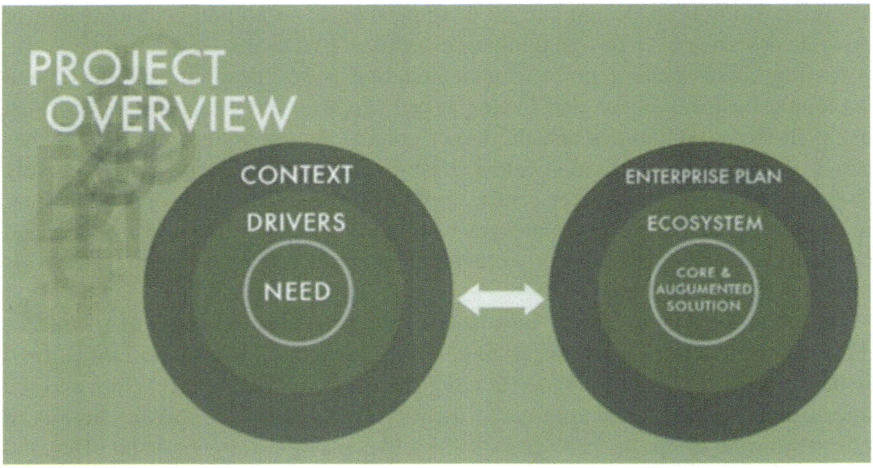

Fig. 6.5a Sustainable product development—Project overview. Adapted from Viswanathan (2016)

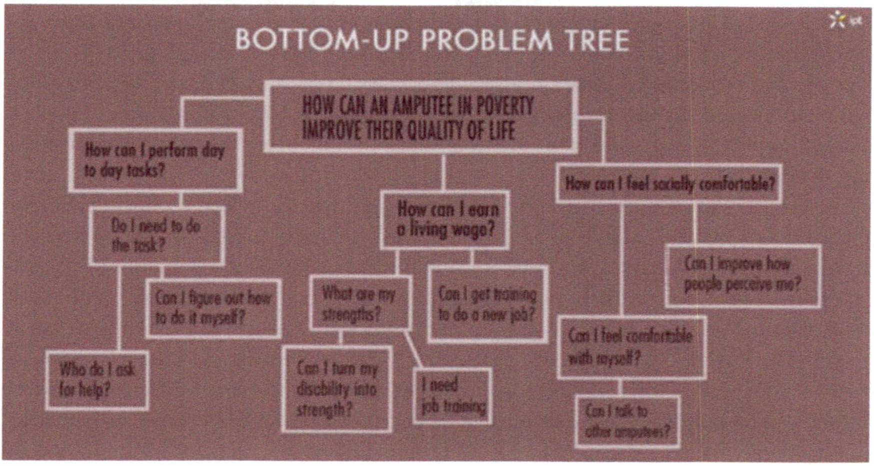

Fig. 6.5b Sustainable product development—Bottom-up need deconstruction. Adapted from Viswanathan (2016)

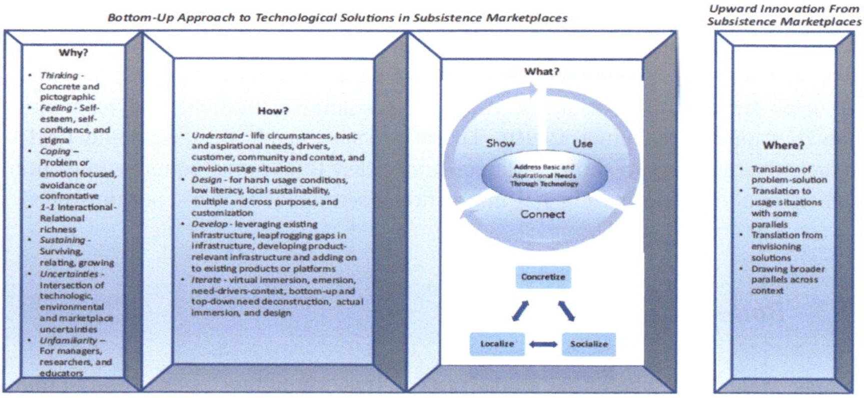

Fig. 6.5c Sustainable product development—Technological solutions in subsistence marketplaces. Adapted from Viswanathan and Sreekumar (2019)

The bottom-up approach in its truest sense is neglected or underemphasized. In this regard, we contend that marketing, despite roots in organizing around the customer, is not bottom-up enough. The bottom-up approach builds from the foundation of need, drivers, consumers, communities, and larger contexts. The top-down approach builds from the foundation of organizational goals, capabilities, and so forth.

In addition to the core solution, we emphasize the ecosystem, which, extrinsic to the product, enables awareness, knowledge, use, and maintenance (Fig. 6.5). Such a focus is important in light of deprivation being on multiple fronts and not just on the product need. The ecosystem, such as local partners who serve as intermediaries for sustainable technologies and products, is vital as a buffer in settings lacking infrastructure, and information, among other factors. As a result, social sustainability may also be enhanced through creating livelihoods for those who add sustainable value. Such an approach also builds on the potential relation strength in 1–1 interactional marketplaces.

We aim to find the meeting point between nascent top-down technologies, and emergent bottom-up needs (Fig. 6.5). Understanding marketplaces and designing and implementing sustainable products and technologies bottom-up enables blending top-down technology with bottom-up culture and social milieu. Thus, sustainable solutions are developed that be shown and used with ease, while enabling people to connect. Sustainable products and technologies represent the *what.* They need to be designed and implemented to address the *how* and *why,* addressing culture that has sustained over a long time. Just as relational or social richness is a double-edged sword, so too are sustainable products and technologies. Bottom-up marketing has the potential to enable such solutions to empower, while enhancing social, environmental, and economic sustainability.

Among the insights generated are product design for local sustainability, harsh usage, low literacy, multiple purposes and customization and product development leveraging current infrastructure, while leapfrogging gaps, or developing product-relevant infrastructure, and adding on to current products or platforms (Viswanathan, 2016; Viswanathan & Sridharan, 2012; Fig. 6.5). Furthermore, "connect," "show," and "use" are central in trial and usage of sustainable products and technology (Viswanathan & Sreekumar, 2019). Therefore, sustainable design should lead to products that are easy to show or demonstrate, easy to use, and enable connections to others in a relationally rich, 1–1 interactional settings (Fig. 6.5—bottom-right).

6.2.2 Bottom-Up Marketing

We have argued elsewhere that marketing practices should be reconceived bottom-up, such as in stitching groups of customers into segments, evolving samples for market research bottom-up, and developing distribution pathways from the ground up (Viswanathan, 2013, 2016). Bottom-up marketing (Viswanathan et al., 2024b) speaks to the organizational level of scaling and growing sustainable solutions bottom-up—the *why* of convergent vision for change balanced with divergent customized purpose across geographies; the *how* of convergent organic growth and evolution balanced with divergent grounded diffusion; and the *what* of convergent baseline platform balance with divergent recreation and adaptation of innovation (Fig. 6.6). Vision at the organizational level is central in addressing sustainability in the twenty-first century. But a bottom-up marketing lens emphasizes the need to reconfigure purpose based on ground-level realities in different geographies. Sustainability being nascent is ideally suited to bottom-up marketing, particularly in subsistence marketplaces and beyond. The confluence of uncertainties along with the confluence of unfamiliarities and the distinctiveness of each geography are compounded in the arena of sustainability, as noted earlier. Thus, organizations should be nimble at all levels in terms of scale and growth, allowing for evolution of purpose while adopting a broad vision (Viswanathan et al., 2024; Fig. 6.6). Indeed, the regional effects of climate change are emergent and overlaid on the uncertainties in subsistence marketplaces. Marketing organizations should be nimble at the level of purpose, process, and substance— Similarly, organic growth may converge across geographies whereas the nature of diffusion may diverge, given ground realities in the multi-faceted domains of subsistence that intersect with sustainability. So too is the "what" or offering in terms of having convergent elements but diverging in being evolved bottom-up.

Fig. 6.6a Top-down versus bottom-up global marketing. Adapted from Viswanathan and Sreekumar (2021)

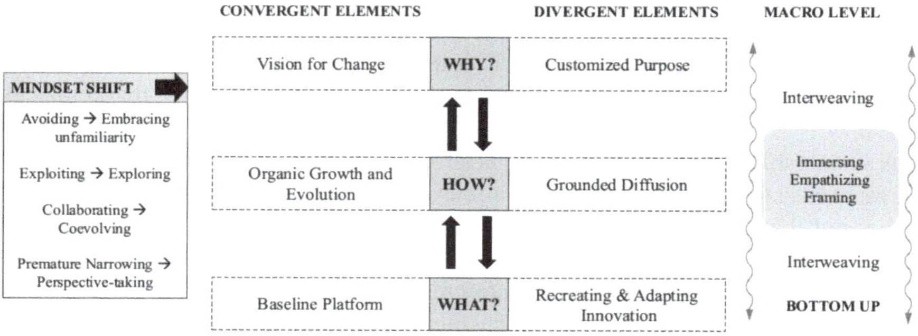

Fig. 6.6b General & granular elements of bottom-up marketing. Adapted from Viswanathan et al. (2024b)

In terms of mindset, bottom-up marketing shifts from avoiding to embracing unfamiliarity, exploiting to exploring, collaborating to coevolving, and premature-narrowing versus perspective-taking. Through immersion, empathizing and frame phases, the bottom-up is interwoven with the macro level.

Customer-driven elements in a top-down approach are not sufficiently bottom-up in our view, nor is contemporary Marketing. We present a complementary counter-perspective, that involves interweaving the bottom-up with the top-down. Thus, it is both grounded in reality and the day-today issues customers face and encompasses a macro-level perspective, traversing different levels of analysis and their interplay. Compartments or silos in marketing based on individual versus higher organizational units of analysis are bridged in this approach. Our approach traverses micro, meso, and macro levels, starting at the micro customer–entrepreneur marketplace level and traversing the path bottom-*up*. Indeed, bottom-up marketing can realize the potential of marketing to be a disciplinary hub that bridges to sustainable development. We argue for an "inside-out" mindset shift, wherein marketplaces are explored and discovered than where preconceptions are applied, or products or processes customized. Such a mindset will be critically important in the nascent and fast-evolving arena of sustainability.

6.2.3 Sustainable Market Orientation

We have argued for a sustainable market orientation derived from three levels of analysis (Fig. 6.7; Viswanathan et al., 2009a, 2009b). The product level blurs the product with betterment of better basic life circumstances. This emphasizes the need to understand deprivation in multiple realms and broader life circumstances when designing sustainable products and business models. Such deprivation is particularly relevant in the domain of the environment. The relationship level blurs the social and the economic, as individuals understand exchange at a 1–1 interactional human level. Thus, fairness and trust are important, all the more so in contexts which face the worst of environmental impacts. Finally, the marketplace level blurs markets with the social milieu. Thus, we argued that going beyond doing good and doing well in parallel, doing good may be essential to doing well in subsistence marketplaces (Fig. 6.7).

A sustainable market orientation argues for inclusion of social good in a product-relevant sense to create sustainable solutions in subsistence marketplaces. The point here is not that a business becomes a charitable organization. Rather, deep understanding of what constitutes individual and community welfare in relation to the business offering provides central insights to create sustainable solutions. In turn, such social good would be incorporated into business processes, outcomes, and evaluations as it is product-relevant being product relevant, social good into the organizational culture. This involves "purposive understanding, addressing customer needs and welfare, and implementing business plans infused with social good" (Viswanathan et al., 2009a, 2009b; Fig. 6.7). Such blurring of social good and business purpose

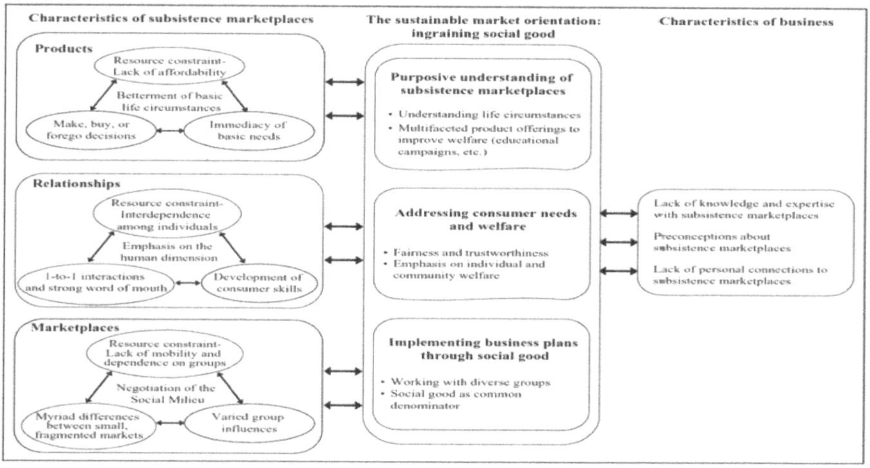

Fig. 6.7a Characteristics of subsistence marketplaces & need for a sustainable market orientation. Adapted from Viswanathan et al. (2009a)

in these marketplaces involves a holistic approach, rather than compartmentalize product offerings from social responsibilities. Indeed, such blurring occurs between diverse types of organizations to work with in order to negotiate the social milieu.

We note how such an orientation is central in the era of sustainability. Products and betterment of life circumstances are indeed blurred at a deeper level with very significant consequences. The social and the economic are blurred as is the environmental in achieving the triple bottom-up line. And markets and the social milieu are blurred, indeed with the environmental as well. Our orientation goes beyond the notion of doing good when doing well to the notion of doing good as being essential to doing well.

6.2.4 Social Enterprise of Marketplace (Marketing) Literacy Encompassing Sustainability Literacy

For addressing sustainability in subsistence marketplaces, another important element is education for those at the bottom of society that enables learning about a very complex set of issues in different domains of subsistence. Relevant, engaging, and personal sustainability education that encompass local and global perspectives can empower people (UNESCO, 2017). Addressing thinking, feeling, and coping constraints, while building on the potential for relational richness, our marketplace literacy education is concretized, localized, and socialized (Viswanathan et al., 2008). In terms of sustainable practices, given the cognitive constraints discussed, know-why is central and not addressed in typical educational programs. Marketplace literacy

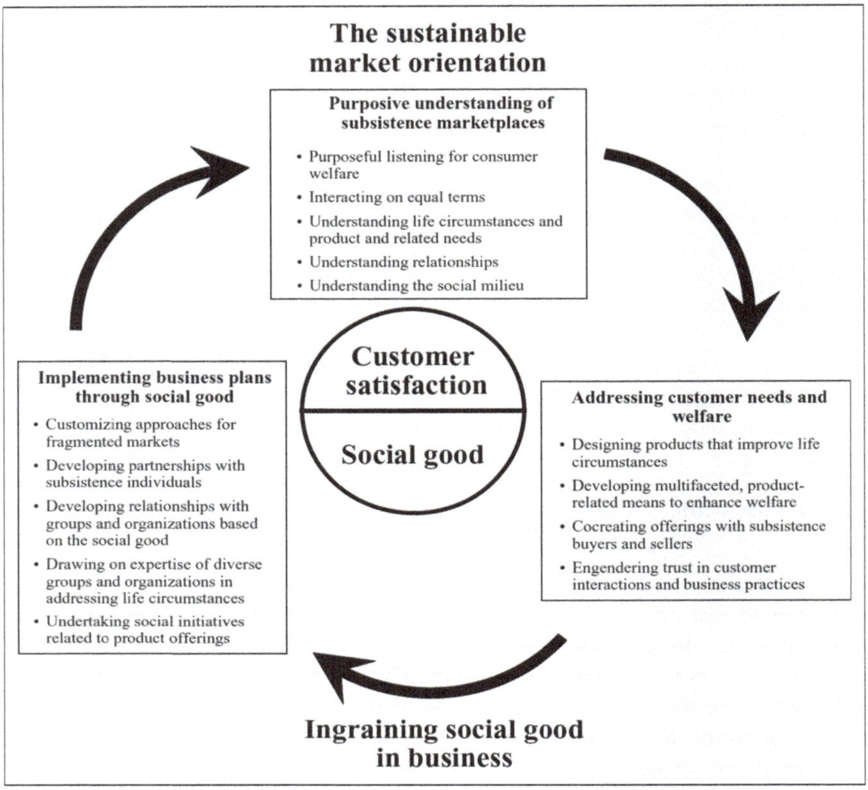

Fig. 6.7b Sustainable market orientation. Adapted from Viswanathan et al. (2009a)

Fig. 6.7c Sustainable market orientation—Doing good—Doing well. Adapted from Viswanathan et al. (2009a)

encompasses knowledge and skills as consumers and entrepreneurs and encompasses consumer, entrepreneurial, and sustainability literacy. It is fundamentally bottom-up and builds off the lived experience of participants, to address the cognitive challenges discussed earlier. The educational content and delivery is concretized, localized and socialized, the latter building of the potential relational richness to create mutual learning that bypasses lack of literacy. Such education has been shown to impact personal wellbeing as well as consumer confidence and decision-making and entrepreneurial intentions and startups (Viswanathan et al., 2021). Such bottom-up education beginning with the know-why can empower individuals and communities in managing environmental issues while enhancing individual and collective efficacy. Central to marketplace literacy is learning how to learn. With a multitude of disruptions due to climate change, such learning is critical to adaptation. Marketplace literacy can be geared toward specific sectors (e.g., agriculture and post-harvest loss prevention) and toward understanding the connections between types of sustainability (e.g., draughts and impact on health and livelihoods).

We note how the lack of socio-political will among the more powerful segments of society places the burden of adaptation on those with the least. We also note the necessary hypocrisy of teaching sustainability literacy where footprints are the least, to address local issues and sustenance. As individuals sustain through survival, relatedness and growth, they face tradeoffs and synergies which can be elucidated through marketplace literacy education.

6.2.5 Symbiotic Academic Social Enterprise

This stream evolved the notion of symbiotic academic social enterprise—understanding the radically different context of subsistence marketplaces, creating marketplace literacy education for such contexts through social enterprise, and creating educational innovations about such contexts geared toward sustainability. As described above, marketplace literacy including sustainability literacy evolved in parallel with the research, as did educational initiatives described below. This social enterprise is highly synergized with research and with educational innovations for university students and managers. For instance, research in parallel led to the intuitive development of marketplace literacy, which is now the focus of ongoing research.

Given the confluence of uncertainties and of unfamiliarities, and the distinctness of each subsistence context, the nascent arena of sustainability in subsistence marketplaces is ideally suited for symbiotic academic-social enterprise (Fig. 6.8). It can be distinguished from action research, coevolving academic and social initiatives through sustained immersion over a long timeframe. Furthermore, rather than a linear relationship between research and practice, each facet enriches the other in a circular relationship, creating mutually enriching loops. As researchers and practitioners are at a confluence of unfamiliarities, such an approach is of particular importance for sustainability. In terms of our overall approach to research, education, and practice, we emphasize holistic, interdisciplinary approaches that examine the interplay between different domains of subsistence and implications for sustainability. For researchers, symbiotic academic-social enterprise combines relevance and rigor which is critically important in this arena.

This essentially bottom-up approach is ideally suited to inform policy and practice for sustainable development as it begins at the level of ground reality and traverses upward in terms of levels of aggregation. Moreover, research and practice are enacted hand-in-hand for purposes of gaining insights. Specific to sustainability in subsistence marketplaces, individuals confront the reality that is, rather than that *can be* if powerful entities and sectors worked toward addressing environmental issues effectively. The highest burden is on those least able due to lack of resources. Enabling those at the bottom of society to understand the projected effects of climate change for specific communities and regions is of paramount importance. Education that evolved bottom-up can be of much value. We demonstrate the evolution of one such educational program elsewhere (Fig. 6.8; Viswanathan et al., 2020). Helping pastoralists understand flood and drought cycles or helping homemakers understand how cooking fuel sources can impact better management of forests. In turn, such education can enable action that benefits communities. Furthermore, bottom-up insights can percolate upward to impact marketing and public policy. Researchers, educators, and practitioners can work together through symbiotic academic-social enterprise, creating mutually enriching loops through sustained immersion in ground realities. In turn subsistence communities strive to do what is in their immediate control.

Another facet of symbiotic academic-social enterprise is in education—through curricular innovations around sustainability and subsistence marketplaces (Viswanathan, 2012; Viswanathan et al., 2011; Fig. 6.8). These educational innovations have been disseminated to educators world-wide (https://www.subsistencemarketplaces.org/education.html—360 videos, day in the life videos, movie, multimedia immersion exercises, qualitative interviews, project reports). We have created

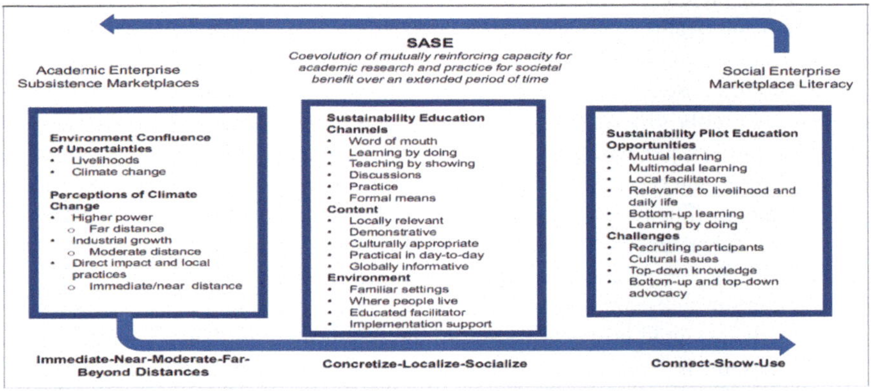

Fig. 6.8 Symbiotic academic social enterprise for subsistence and sustainability. Adapted from Viswanathan et al. (2020)

learning experiences on sustainability from first-year undergraduate students to executive and doctoral levels and practitioners such as UNHCR staff. These experiences focus substantively on subsistence marketplaces and methodologically on a bottom-up approach, working toward the large goal of sustainable development (Fig. 6.9). They capture the essence of the bottom-up approach and have led to unique processes for sustainable product design and development, as discussed earlier. About 120–150 group projects a year for first-year business students over 15 years ending with sustainable business plans and 55 yearlong group projects over 13 years with detailed product prototyping and comprehensive business plans are among the outcomes. Each of these projects emphasizes a bottom-up approach beginning with immersion to emphasize understanding of the larger context and drivers in addition to needs, where environmental issues in subsistence marketplaces are front and center. The steps in our sustainable business plan show how Marketing is front and center (Fig. 6.9). Immersive exercises including virtual interviews as well as designing and communicating the value proposition in light of audiences with low literacy, and positioning in relation to the competition which can include doing nothing or making instead of buying are very much in the domain of Marketing.

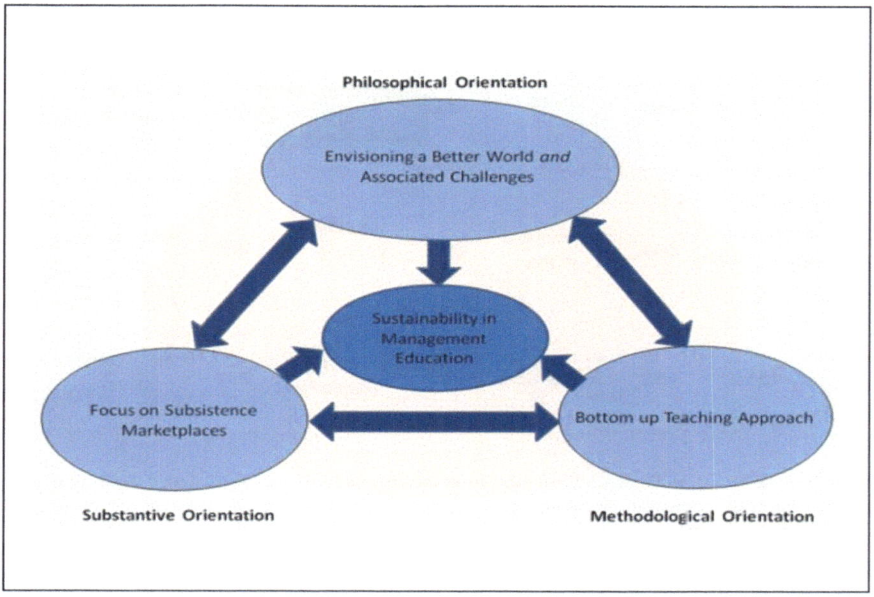

Fig. 6.9a Marketing and management education on sustainability and subsistence marketplaces—Philosophical, substantive & methodological orientation. Adapted from Viswanathan (2012)

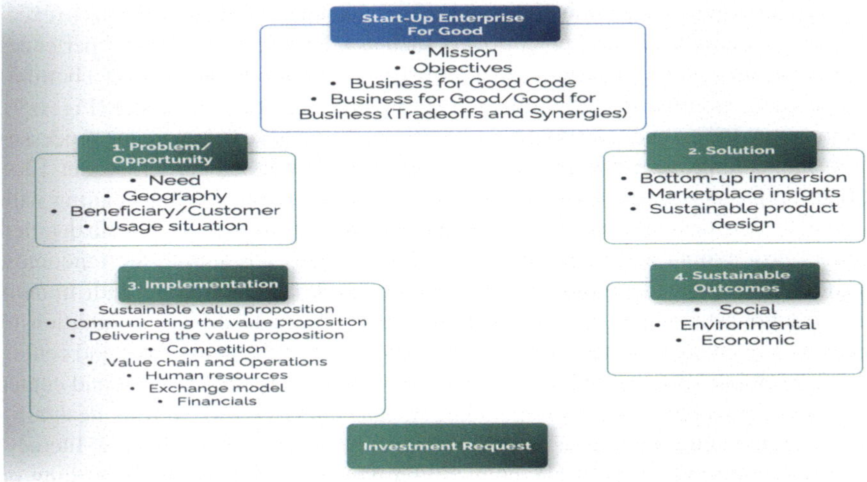

Fig. 6.9b Marketing and management education on sustainability and subsistence marketplaces—Sample project outline

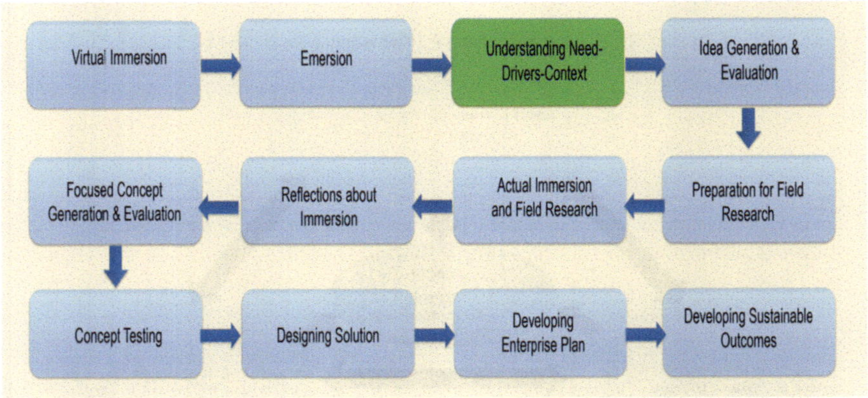

Fig. 6.9c Marketing and management education on sustainability and subsistence marketplaces—Sample project phases in bottom-up immersion, emersion, design, innovation & enterprise. Adapted from Viswanathan (2016)

6.3 Discussion

Subsistence marketplaces are most affected by environmental issues, yet least able to address them in terms of material resources. Drawing from more than two and a half decades of work with low-literate, low-income consumers, entrepreneurs, and marketplaces—i.e., subsistence marketplaces, we presented key insights at the intersection of sustainability and subsistence. Our approach is bottom-up, beginning at the micro-level behavioral understanding of subsistence marketplaces as they intersect with the environment, consumption, and conservation. This contrasts with the customary vantage point in sustainability of being top-down. We find synergies between research, education, and practice, through symbiotic academic-social enterprise. The interplay between subsistence and sustainability using a bottom-up vantage point generates new insights. We cover understanding of subsistence marketplaces and sustainability, and addressing challenges through product design and development, business models, education, academic-social enterprise, and sustainable development goals.

To work toward social and environmental sustainability along with economic sustainability in subsistence marketplaces is daunting and challenging. Yet, among the biggest sources of optimism are the resilience, aspirations, and actions of those living in these communities as they survive, relate, and grow—i.e., strive to sustain in day-to-day reality. Understanding and addressing this reality bottom-up is central to sustainability from a marketing perspective.

Noteworthy is the intersection of this stream with SDGs, which encompass realms ranging from the natural environment and ecology (Wood & DeClerck, 2015), to human well-being (Buse & Hawkes, 2015) and technology (Wu et al., 2018). SDGs emphasize balancing economic growth with human needs and environmental concerns, representing a network of phenomena (Le Blanc, 2015). Social vantage points are central to development through SDGs (Briant Carant, 2017), emphasizing caution against purely top-down approaches (Hajer et al., 2015). Rather, stakeholder needs at multiple levels need to be addressed (Rasul, 2016). We privilege voices from subsistence marketplaces noting the intersection between SDGs and domains of subsistence—health, nutrition, water, energy, etc. Of particular relevance are SDGs related to basic physical needs. Challenges and trade-offs across domains of subsistence can be understood through the bottom-up approach beginning at the micro level. Thus, our approach is well suited for SDGs rooted at the micro level (Fig. 6.10). For SDGs at a more macro level, the bottom-up approach provides insights and participatory voice for subsistence individuals—typically most affected, with least say.

SDGs emphasize livelihoods and economics. Our perspective brings into focus buyers, sellers, and beyond. In this regard, marketplace literacy focuses on developing know—why covering consumer, entrepreneurial, and sustainability literacy. Livelihoods are "a means of support or subsistence" ("securing the necessities of life" or "a way of earning money in order to live" (http://www.merriam-webster.com/dictionary/livelihood). The latter definition is narrower than the former in that the means of support can extend beyond money. Another definition is that livelihood is "a means of supporting one's existence, especially financially or vocationally" (http://dictionary.reference.com/browse/livelihood), emphasizing specific means respectively—monetary and through occupation. The "means of support or subsistence" arise from a variety of sources—making rather than buying, apportioning consumption, and conserving resources in the process of being a consumer. Being creative in using the resources around, whether it be firewood or water is a central aspect of "livelihood"

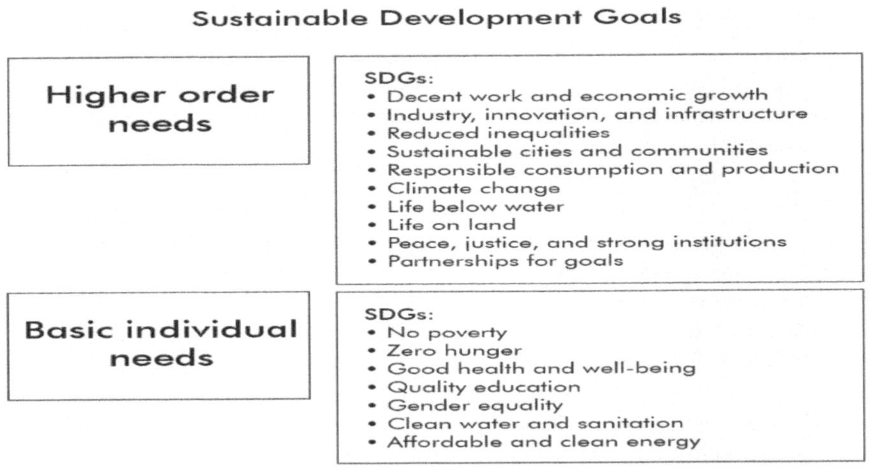

Fig. 6.10 Sustainable development goals

in the broader sense. Making rather than buying is another aspect, be it health-related treatment, furniture, or food ingredients. Producing partially for one's own consumption is another aspect as well, involving a vocation (say, agriculture) that is used to address a basic need. And finally, conserving resources by being a better customer is another aspect, leaving individuals with better products and money in hand. All these aspects represent "means of support or subsistence". Thus, we assert that the elucidation of SDGs should emphasize marketplaces to a greater extent.

The implications of our work for sustainability and marketing in subsistence marketplaces are summarized in Fig. 6.11. Implications encompass understanding exchanges for their own sake across levels of socio-economic status, moving beyond needs and wants to aspirations, considering the dual roles of the customer-entrepreneur, and going beyond considering not only the customer but the community and larger context. Implications cover conducting market research through virtual/actual immersion and emersion, designing sustainable product design through the bottom-up approach for unanticipatable usage situations, and considering upward or reverse innovation *from* subsistence marketplaces to other contexts. In terms of education, we argue for moving from compartmentalized marketing education to integrative learning experiences on sustainability. We argue for inter-disciplinary as well as intersector approaches, going as far as to propose academic-social enterprise, a parallel, mutually enriching relationship between research, (education), and practice. For communities, a central implication is creating and delivering bottom-up marketplace literacy (marketing) including sustainability literacy education. This can be done for different sectors such as agriculture and can emphasize social and environmental sustainability in addition to economic sustainability. As climate change leads to a variety of disruptions, such education that is bottom-up and focusing on know-why can empower individuals to learn how to learn and adapt. Finally, we demonstrate how doing good can be essential to doing well, for particular importance when considering the triple bottom line.

Limitations and pitfalls are ever-present in working at the intersection of subsistence and sustainability. In addition to subsistence contexts being distinctive, they encompass a wide range of settings such as refugee settlements, war-torn areas, and locations facing natural disasters. For instance, relational or social richness can be a foundation in sustainable development through marketing but may be disrupted due to communal conflict or due to political control over people getting together or collective action. Social good as a common denominator is central in enacting a sustainable market orientation but, of course, vulnerable to corrupt practices and political influences. Pitfalls of the bottom-up approach relate to the need to carefully interweave it with the top-down. Aggregating insights to a usable form is central to being bottom-*up*. Our approach has aggregated the essence of the micro-level insights to product development and business models. This requires traversing from micro to meso, and macro levels. The bottom-up approach is very challenging as

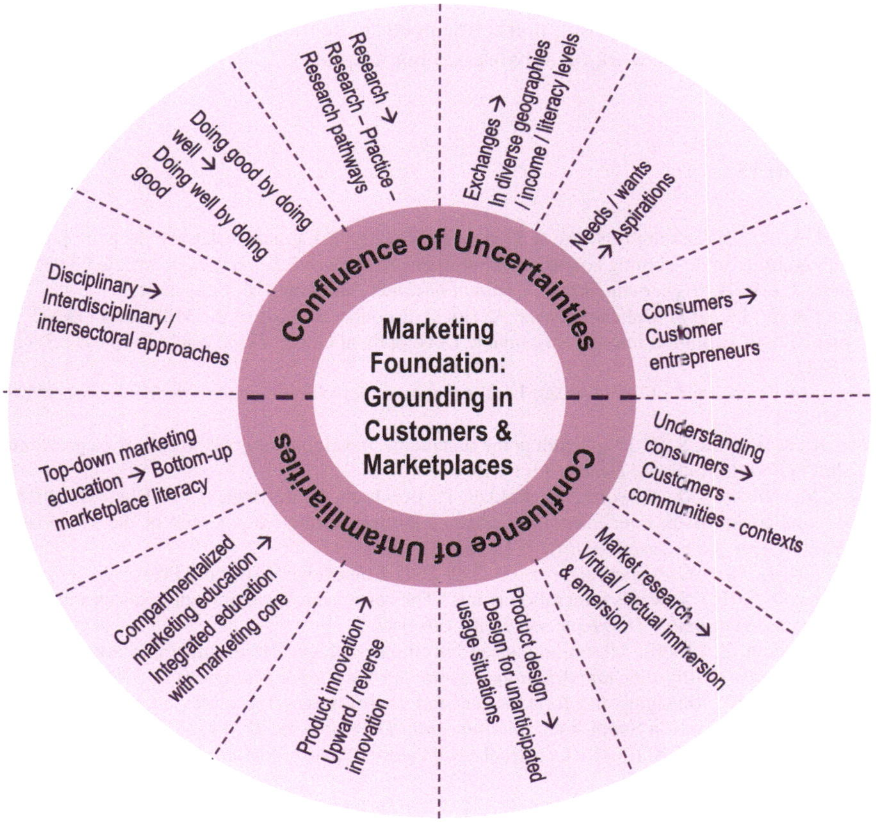

Fig. 6.11 Marketing advances in an era of disruptions. Adapted from Viswanathan (2019)

our default or fallback option is to be top-down, whether at the individual level with what we already know or the organizational level in how we view the world from our perspective with preexisting structures and processes. Being bottom-up requires the reality outside at a micro-level to percolate into our thinking. This goes beyond horizontal influences from other organizations or individuals to extending vertically to different socio-economic levels. Means for aggregating insights upward are central to converting knowledge of ground reality into actions. Another pitfall here is being "lost in the weeds" or overreacting to idiosyncratic ground reality, wherein the interplay or interweaving with the top-down becomes crucial. We have developed several bottom-up processes and detailed examples from our experiences, such as with marketplace literacy.

In conclusion, the bottom-up approach is much needed though often neglected, in understanding and addressing sustainability in subsistence marketplaces. Our bottom-up approach demonstrates a bridge between the discipline of marketing and sustainable development, ideally positioned to understand and address the micro-level reality of the customer-beneficiary. Crucial to such endeavors is the disciplinary lens of marketing with its focus on the customer and the marketplace. However, to create a greater positive impact, marketing needs to be more bottom-up, as illustrated through the stream of work on subsistence marketplaces.

References

Ahmed, S. A., Diffenbaugh, N. S., & Hertel, T. W. (2009). Climate volatility deepens poverty vulnerability in developing countries. *Environmental Research Letters, 4*(3), Article 034004.

Barnett, R. (1997). *Higher education: A critical business.* McGraw-Hill Education.

Briant Carant, J. (2017). Unheard voices: A critical discourse analysis of the Millennium Development Goals' evolution into the Sustainable Development Goals. *Third World Quarterly, 38*(1), 16–41.

Brown, M. E., & Funk, C. C. (2008). Food security under climate change. *Science, 319*(5863), 580–581.

Buse, K., & Hawkes, S. (2015). Health in the sustainable development goals: Ready for a paradigm shift? *Globalization and Health, 11*(1), 13.

Hajer, M., Nilsson, M., Raworth, K., Bakker, P., Berkhout, F., de Boer, Y., & Kok, M. (2015). Beyond cockpit-ism: Four insights to enhance the transformative potential of the sustainable development goals. *Sustainability, 7*(2), 1651–1660.

IPCC. (2014). *Fifth Assessment Report* (pp. 229–269). Cambridge University Press.

Le Blanc, D. (2015). Towards integration at last? The sustainable development goals as a network of targets. *Sustainable Development, 23*(3), 176–187.

Meze-Hausken, E. (2000). Migration caused by climate change: How vulnerable are people in dryland areas? *Mitigation and Adaptation Strategies for Global Change, 5*(4), 379–406.

Rasul, G. (2016). Managing the food, water, and energy nexus for achieving the Sustainable Development Goals in South Asia. *Environmental Development, 18*, 14–25.

Trope, Y., & Liberman, N. (2010). Construal-level theory of psychological distance. *Psychological Review, 117*(2), 440–463.

UNESCO. (2017). *The United Nations World Water Development Report 2017: Wastewater, the untapped resource.* UNESCO CLD.

Viswanathan, M. (2012). Curricular innovations on sustainability and subsistence marketplaces: Philosophical, substantive and methodological orientations. *Journal of Management Education, 36*, 389–427.

Viswanathan, M. (2013). *Subsistence marketplaces*.

Viswanathan, M. (2016). *Bottom-up enterprise: Insights from subsistence marketplaces*. eBook-partnership, eText, and Stipes Publishing.

Viswanathan, M. (2019). Envisioning marketing advances in an era of disruptions: A bottom-up perspective from subsistence marketplaces. In *Handbook of advances in marketing in an era of disruptions: Essays in honor of Jagdish N. Sheth*. Sage.

Viswanathan, M., Baskentli, S., Gallage, S., Grigortsuk, M., Martin, D., & Subrahmanyan, S. (2020). A demonstration of symbiotic academic-social enterprise in subsistence marketplaces: Researching and designing customized sustainability literacy education in Tanzania. *Journal of Public Policy and Marketing, 40*(2), 245–261. https://doi.org/10.1177/0743915620968506

Viswanathan, M., Chase, L. J., & Jones, M. (2024a). I did not think of myself as a 'customer': The confluence of intertwined vulnerabilities among subsistence consumers through marketplace literacy. *Review of Marketing Research, The Vulnerable Consumer, 21*, 133–152.

Viswanathan, M., Gajendiran, S., & Venkatesan, R. (2008). *Enabling consumer and entrepreneurial literacy in subsistence marketplaces*. Springer.

Viswanathan, M., Jung, K., Venugopal, S., Minefee, I., & Jung, I. W. (2014). Subsistence and sustainability: From micro-level behavioral insights to macro level implications on consumption and the environment. *Journal of Macromarketing, 34*(1), 8–27.

Viswanathan, M., & Rosa, J. (2007). Product and market development for subsistence marketplaces: Consumption and entrepreneurship beyond literacy and resource barriers. In J. Rosa & M. Viswanathan (Eds.), *Product and market development for subsistence marketplaces: Consumption and entrepreneurship beyond literacy and resource barriers*, Advances in International Management Series, Joseph Cheng and Michael Hitt, Series Editors (pp. 1–17). Elsevier.

Viswanathan, M., Rosa, J. A., & Harris, J. E. (2005). Decision making and coping of functionally illiterate consumers and some implications for marketing management. *Journal of Marketing, 69*(1), 15–31.

Viswanathan, M., Rosa, J. A., & Ruth, J. A. (2010). Exchanges in marketing systems: The case of subsistence consumer–merchants in Chennai, India. *Journal of Marketing, 74*(3), 1–17.

Viswanathan, M., Seth, A., Gau, R., & Chaturvedi, A. (2009a). Internalizing social good into business processes in subsistence marketplaces: The sustainable market orientation. *Journal of Macromarketing, 29*, 406–425.

Viswanathan, M., & Sreekumar, A. (2019). Consumers and technology in a changing world: The perspective from subsistence marketplaces (invited). *European Journal of Marketing, 53*(6), 1254–1274.

Viswanathan, M., & Sreekumar, A. (2021). How global marketing can be more global and more marketing: A bottom-up perspective from subsistence marketplaces. *Journal of Global Marketing, 34*(4), 282–291.

Viswanathan, M., Sreekumar, A., Sridharan, S., & Sinha, G. (2024b). Addressing the grand challenge of poverty and inequality through the bottom-up marketing approach: Lessons from subsistence marketplaces and marketplace literacy. *Journal of the Academy of Marketing Science, 52*, 1279–1300. https://doi.org/10.1007/s11747-024-01022-z

Viswanathan, M., & Sridharan, S. (2012). Product development for the BoP: Insights on concept and prototype development from university-based student projects in India. *Journal of Product Innovation Management, 29*(1), 52–69.

Viswanathan, M., Sridharan, S., Gau, R., & Ritchie, R. (2009b). Designing marketplace literacy education in resource-constrained contexts: Implications for public policy and marketing. *Journal of Public Policy & Marketing, 28*(1), 85–94.

Viswanathan, M., Sridharan, S., Ritchie, R., Venugopal, S., & Jung, K. (2012). Marketing interactions in subsistence marketplaces: A bottom-up approach to designing public policy. *Journal of Public Policy & Marketing, 31*(2), 159–177.

Viswanathan, M., Umashankar, N., Sreekumar, A., & Goreczny, A. (2021). Marketplace literacy as a pathway to a better world: Evidence from field experiments in low-access subsistence marketplaces. *Journal of Marketing, 85*(3), 113–129. https://doi.org/10.1177/0022242921998385
Viswanathan, M., & Venugopal, S. (2015). Subsistence marketplaces: looking back, looking forward. *Journal of Public Policy & Marketing, 34*(2), 228-234.
Viswanathan, M., Yassine, A., & Clarke, J. (2011). Sustainable product and market development for subsistence marketplaces: Creating educational initiatives in radically challenging contexts. *Journal of Product Innovation Management, 28*, 558–569.
Wood, S. L., & DeClerck, F. (2015). Ecosystems and human well-being in the Sustainable Development Goals. *Frontiers in Ecology and the Environment, 13*(3), 123–123.
Wu, J., Guo, S., Huang, H., Liu, W., & Xiang, Y. (2018). Information and communications technologies for sustainable development goals: State-of-the-art, needs and perspectives. *IEEE Communications Surveys & Tutorials, 20*(3), 2389–2406.

Madhu Viswanathan is Professor of Marketing in the College of Business Administration at Loyola Marymount University, Los Angeles, CA, and Professor Emeritus at the Gies College of Business at the University of Illinois, Urbana-Champaign, IL, USA. His research programs are on measurement and subsistence marketplaces. He has also authored several books in this research area, including Measurement Error and Research Design (2005), Enabling Consumer and Entrepreneurial Literacy in Subsistence Marketplaces (2008), Subsistence Marketplaces (2013), and Bottom-up Enterprise (2016). He pioneered research in the area of subsistence marketplaces, a bottom-up approach to poverty and marketplaces (www.business.illinois.edu/subsistence). He teaches courses on research methods, subsistence, and sustainability. He founded and directs the Marketplace Literacy Project (www.marketplaceliteracy.org), a pioneering marketplace literacy education platform that has reached more than 100,000 women across four continents. He has received numerous awards and served on the Livelihoods Advisory Board of UNHCR.

Chapter 7
Empowering the Bottom of the Pyramid: Harnessing Innovation for Sustainable Growth

Shaphali Gupta and Rushabh Trivedi

Abstract Billions of individuals worldwide subsist at the Bottom of the Pyramid (BOP), contending with scarce resources and restricted access to vital services such as healthcare, education, and financial facilities. Despite their significant number, this segment often goes underserved and overlooked. This chapter explores how various BOP-focused innovations, such as grassroots innovation, reverse innovation, and Jugaad innovation, can empower the BOP with social, economic, and environmental sustainability. Drawing on theories such as the resource-based view, community-based theory, theory of disruptive innovations, and effectuation theory, this discussion elucidates how each form of innovation can enhance the lives of individuals in low-income settings. First, the chapter highlights social innovation, which addresses the needs of the BOP through the coordinated engagement of multiple stakeholders, including local communities, governmental bodies, and private enterprises. Next, it focuses on grassroots innovation, which emerges within communities to leverage local insights and indigenous knowledge, yielding culturally and ecologically relevant solutions. The discussion then shifts to reverse innovations tailored to provide solutions for emerging markets before being introduced to developed economies. The chapter then examines frugal innovation, which capitalizes on resource efficiency and cost-effectiveness to produce affordable, high-impact solutions that tackle pressing challenges. Finally, it emphasizes jugaad innovation, often driven by improvisation, flexibility, and ingenuity with minimal resources, and embodies the creative resilience inherent in resource-constrained settings. In the end, supported by real-world examples, the chapter draws attention to the role of an enabling ecosystem that includes supportive government policies, accessible infrastructure, access to finance, robust social networks, and technology transfer and how these enablers can influence the BOP through expediting innovation processes and foster sustainable and inclusive growth.

S. Gupta (✉)
Indian Institute of Management, Kozhikode, Kozhikode, India
e-mail: shaphali.gupta@iimk.ac.in

R. Trivedi
MICA, Ahmedabad, India

R. Singh and T. A. Wani (eds.), *Marketing to the Base of the Pyramid*,
https://doi.org/10.1007/978-981-96-7105-2_7

Keywords Bottom of the pyramid · Social innovation · Grassroots innovation · Reverse innovation · Frugal innovation · Jugaad innovation

This chapter addresses the following questions:

1. What is the role of innovations in empowering the Bottom of the Pyramid?
2. Discuss various types of BOP-focused innovations and how they are defined.
3. How do different types of BOP-focused innovation—reverse innovation, frugal innovation, social innovation, grassroots innovation, and jugaad innovation empower the BOP segment to achieve economic, social, and environmental sustainability?
4. What factors influence the successful implementation and scaling of innovations within BOP markets?

Aditya, India's first solar-powered ferry, is a significant advancement that aids BOP communities in Kerala by providing economical and environmentally sustainable transportation. Entirely powered by solar energy, Aditya operates with zero emissions, considerably diminishing its environmental impact and lowering operational expenses relative to conventional diesel ferries. The cost-effectiveness of this innovation significantly improves accessibility for low-income passengers, facilitating access to their employment, education, healthcare, and other essential needs. The ferry promotes local tourism, creates employment opportunities, and replicates sustainable waterways. As a receiver of the 2020 Gustave Trouve Award for the World's Best Electric Ferry, Aditya demonstrates how solar technology can effectively empower bottom-of-the-pyramid populations by enhancing mobility, mitigating pollution, and fostering economic opportunities, thus promoting sustainability (Varma, 2020).

7.1 Introduction

In the modern global economy, a stark disparity in wealth distribution persists, with billions of individuals residing at the bottom of the pyramid (BOP). The BOP is the largest but the poorest social and economic stratum, which includes about 2.7 billion people in the world with an income averaging around \$2 to \$8 per day (Baishya & Samalia, 2020; Prahalad, 2012; Purohit et al., 2021; Rahman et al., 2018; Srivastava et al., 2020). This massive segment represents people of different cultures and ethnic backgrounds with varied needs. Despite such diversity, people under the BOP umbrella lack basic lifestyle facilities, such as access to clean water and sanitation, impaired healthcare services, low-quality food and education, and unavailability of inclusive financial services.

The BOP segment is not only a passive recipient of aid but rather is capable, ambitious, and has the potential to contribute to economic growth. Nevertheless, various multinational and national organizations and their proposed innovation have paid little heed to this population segment, targeting the markets with higher purchasing power and well-structured systems instead (Gupta, 2020). Most organizations view the BOP market as unattractive due to the low individual income, high costs of transactions, and infrastructural constraints such as poor transport networks. Leaving out individuals falling into the BOP may lead to organizations failing to capitalize on the combined buying capacity of the BOP consumers even though they are constrained in their ability to make purchases (Gupta, 2020).

The BOP forms a big part of the world's population and exhibits a vast market with enormous potential and requirements. However, addressing the need of this segment requires a paradigm shift from seeing the BOP as a target for low-cost products to seeing them as part of the economy with its potential to provide economies of scale. If businesses combine the social profit motive with various innovations targeting BOP consumers, it could help address multiple problems in serving this segment and fulfilling its needs. The social profit orientation pertains to business strategies synchronized to create financial returns and social benefits by combining profit with social purpose (Berry et al., 2024).

This chapter contends that innovation is a pivotal tool for empowering the BOP. By examining five specific categories of BOP-oriented innovation: social innovation, grassroots innovation, reverse innovation, frugal innovation, and jugaad innovation, the chapter illustrates how these innovative approaches can deliver affordable, accessible, and sustainable solutions tailored to low-income communities. Real-world examples demonstrate how innovation can enhance access to essential services, create economic opportunities, and improve the overall quality of life for BOP populations, thereby unlocking their potential to contribute meaningfully to economic growth. In line with the United Nations Sustainable Development Goals (SDGs), the chapter delineates the extent to which innovative strategies can be effectively implemented to gain social, economic, and environmental outcomes that benefit the BOP communities. Lastly, the chapter highlights the enablers and contexts that expedite and support the process of innovations directed toward BOP (Fig. 7.1). The insights presented in this chapter may appeal to entrepreneurs, governments, non-governmental organizations, and other stakeholders committed to sustainable development.

7.2 Innovation and Bop Empowerment

7.2.1 Social Innovation

A collaborative effort among governments, for-profit entities, and local communities is essential to pursue the ambitious target of achieving SDGs (Aksoy et al., 2019). In doing so, social innovation is a progressive process that aims to create new products, services, and systems to meet the needs of the BOP. In its simplest form, social innovation is the creation of new ideas and solutions for a society where the involvement of various actors, including the public, private, and non-governmental sectors, is imperative (Gupta et al., 2020). Social innovation is a process that enables communities to engage actively in finding novel and long-term solutions that are culturally appropriate and sustainable, using local knowledge and expertise (Van Oost et al., 2009).

In the context of social innovation, the resource-based view (RBV) can profoundly delineate the extent to which social enterprises can succeed in generating significant value (Barney, 1991). The key lies in how these entities strategically leverage and transform local resources into viable solutions for community problems. The RBV theory emphasizes the strategic utilization of resources by leveraging unique and valuable capabilities to establish a competitive advantage. Effectively mobilizing social, human, and community resources can contribute to achieving social impact in underserved markets, such as the BOP. For example, Grameen Bank (a bank for the rural segment in Bangladesh) leverages social capital and social trust, allowing it to create a sustainable microcredit model for the rural segment. This bank's innovative use of intangible assets successfully transformed financial exclusion into economic empowerment for the BOP segment.

Example: Grameen Bank
In Bangladesh, poverty is compounded with no access to credit, especially for women. The conventional financial institutions did not offer credit to people who could not provide collateral, which meant that the poor remained poor as they could not escape the poverty bracket. This financial exclusion prevented the development of the economy and maintained social injustice.

Influenced by the Bangladesh famine of 1974, Muhammad Yunus founded the Grameen Bank, an initiative that offered microcredit through small loans to cater to the needs of disadvantaged groups (Forum, 2024). The bank employed the group lending model, indicating heightened community support, which resulted in lowered default rates. The Grameen Bank worked based on trust and social capital instead of collateral, allowing poor people to borrow money to uplift their lives. The Grameen Bank, with the microcredit model, revolutionized the BOP marketplace. It empowered millions to engage in small-scale businesses to earn more and promote their economic development. Such an approach in the financial domain has gained global acceptance due to its financial viability, leading to advances in education, health, and living conditions, thereby facilitating overall social development, particularly in developing and underdeveloped countries.

Offering financial services to the impoverished without collateral allowed Grameen Bank's microcredit approach to showcase effective, sustainable practices for the BOP (Prahalad, 2012). This strategy encouraged financial progress and an equitable society by endorsing sustainability in inclusive capitalism (Hart, 2010). The bank's achievements demonstrated how microfinance can support sustainable development and promote better living standards, thus empowering people with low incomes.

Within the RBV framework, the Grameen Bank recognized its social capital and community resources as essential strategic assets. The bank utilized the intangible asset of trust rather than traditional collateral, facilitating credit access for underprivileged populations. The group-lending strategy of Grameen Bank leveraged community support and social cohesion as essential resources, promoting collective responsibility and dedication among borrowers (Tiwari, 2017).

7.2.2 Grassroots Innovation

Grassroots innovation (GRI) can be defined as "innovations originating from economically disadvantaged people, in an informal market ecosystem, creatively deploying their native skills and local knowledge to resolve their community problems in an affordable and sustainable way" (Gupta, 2020, p. 948). GRI refers to products developed by individuals, termed grassroots innovators, who belong to economically weaker sections, possess a deep understanding of their community's needs and challenges, are aware of their resource limitations, and are intrinsically driven to address community issues through innovative solutions that integrate local, contextual, and traditional knowledge (Gupta, 2020). In the BOP markets, the right products can yield several benefits, such as high productivity, reduced illness and better health, low rate of accidents, and increased environmental sustainability (Seyfang & Smith, 2007). Grassroots innovators typically hail from the BOP and leverage local knowledge and expertise to create affordable goods and services that address diverse issues without institutional support (Kumar & Bhaduri, 2014). However, broader descriptions also suggest the role of external assistance through lending expertise and/or collaborations (Raj et al., 2022). GRIs emerge when innovators are motivated by personal incentives to contribute to society, are committed to preserving traditional knowledge, or are involved in a hobby and/or seek financial benefits (Nair et al., 2017). GRIs can generate commercial and non-commercial value for its stakeholders in several ways. For instance, the commercial value consists of benefits such as increased revenue and patents, local market penetration, and new business and partnership opportunities. The non-commercial value entails societal and environmental benefits such as empowerment of local communities, preservation of traditional knowledge, skill development, and biodiversity conservation (S. Gupta, 2020). In sum, grassroots innovators have the potential to generate economic value, foster knowledge, and create opportunities for the benefit of society (Appadurai, 2000).

In the GRI context, the community-based theory elucidates ideas about creating and implementing effective programs in a community and addressing community needs (Peredo & Chrisman, 2006). Also, GRI innovation encourages the local community's and individuals' proactive involvement in cooperating to devise solutions tailored to community needs and circumstances. The origin of GRI generally does not arise from formal R&D environments but rather from individuals or networks inside marginalized populations who possess a deep understanding of local issues. Grassroots innovators can develop sustainable, culturally relevant, economically viable, and scalable solutions by leveraging indigenous knowledge, traditional techniques, and community engagement.

Example: The Sanitary Pad Revolution
Arunachalam Muruganantham, a resident of Coimbatore, India, exemplified grassroots innovation propelled by a profound understanding of local requirements. After recognizing the use of unsanitary cloth by local women in the village during menstruation, Arunachalam Muruganantham, also known as 'Padman', created a machine that was cost-effective, compact, and produced sanitary pads at an affordable price compared to other commercial players in the market. By locally sourcing materials and ensuring that the sanitary pads were economical, safe, and comfortable, Muruganantham's concept not only rendered sanitary pads accessible to millions but also generated business opportunities for rural women, enhancing their social status and dismantling the taboo around menstruation. Muruganantham's example highlights that grassroots innovation encompasses solutions that facilitate social change, empowering underprivileged groups and enabling the BOP to live a better life. Padman's example demonstrates the importance of grassroots innovation and local communities' role in fostering innovation to address specific challenges (Seyfang & Smith, 2007; Smith et al., 2014).

7.2.3 Reverse Innovation

Reverse innovation is defined as "clean slate, super value products that are technologically advanced created to meet the unique needs of relevant segments, initially adopting in the emerging markets followed by the developed countries" (Malodia et al., 2020, p. 1010). Reverse innovation denotes creating products and services tailored to emerging markets and launching them in developed markets. Contrary to the innovation models of building products and services that primarily cater to the needs of individuals in developed economies, reverse innovations initially permeate emerging markets and are then pushed into the developed markets. The principles of reverse innovation are based on cost optimization, flexibility, non-complexity, and replicability (Govindarajan & Ramamurti, 2011). Entities involved in reverse innovation aim to develop affordable products and services without sacrificing essential features and characteristics. The basic concept is to keep the products simple and provide intuitive interfaces without extensive training or detailed knowledge.

The disruptive innovation theory is particularly relevant in reverse innovation, as it explains how innovations originating in emerging markets can redefine established industries in developed markets (Markides, 2006). The theory contends that disruptive innovations begin with a smaller set of customers or a new market, initially with simpler, low-cost products that provide sufficient utility to gain traction in unattractive market domains. While traditional companies focus on refining high-end products for the mainstream market, reverse innovations succeed by meeting the most basic needs of the underserved and then propagating in the mainstream markets. By utilizing this dynamic, innovations created for resource-constrained environments can challenge existing solutions by demonstrating the possibility of reverse innovation disrupting global industries.

Example: GE's Portable Ultrasound Machine
General Electric (GE), a leading multinational conglomerate, found a critical gap in the overcrowded medical imaging market to serve remote areas in China and India. The existing models of ultrasound machines were costly and bulky. They required a constant source of electricity and skilled personnel who were not readily available in rural areas, particularly in the developing markets. Thus, the General Electric team in China created a battery-operated portable ultrasound machine that was cheaper than its counterparts. This innovation allows rural healthcare givers to address the needs of that market efficiently and conveniently. GE redesigned the technology and narrowed the features, removing the complexities that increase costs and decrease usability in a resource-limited environment (D'Monte, 2016).

The disruptive portable ultrasound machine turned out to be a revolutionary product in the healthcare delivery system in rural regions, facilitating the provision of diagnostic services to reach remote areas. The lowered cost of medical imaging makes it affordable for patients with little or no income and helps in early diagnosis and intervention, reducing mortality rates. GE identified the use of portable ultrasound in developed markets as a business opportunity and then launched the device in the United States. It proved highly valuable in emergency departments, ambulances, and sports medicine, where mobility and rapid results were essential. The device's success in affluent markets highlighted the opportunity for reverse innovation for many consumers (Govindarajan & Trimble, 2012). GE's design of a low-cost ultrasound system exemplifies disruptive innovation focused on social change through more accessible and available technologies that open new markets (Markides, 2006).

7.2.4 *Frugal Innovation*

Frugal innovations can be defined as "a resource-scarce solution (i.e., product, service, process, or business model) that is designed and implemented despite financial, technological, material or other resource constraints, whereby the outcome is significantly cheaper than competitive offerings (if available) and is good enough to meet the basic needs of customers who would otherwise remain un(der)served" (Hossain et al., 2016). Frugal innovations are often linked to sustainable solutions due to their minimized use of resources, affordable pricing, and availability for the masses (Albert, 2019). Such innovations rely on cost-effectiveness and resource efficiency, making them available, especially for the BOP. To this end, frugal innovations exemplify value propositions by using affordable materials and operational processes that make products and services as good as those in the conventional market but at a fraction of the price (Zeschky et al., 2011). By focusing on product effectiveness, dependability, and user-oriented design, frugal innovations can enable the development of solution-centric products for the masses (Hossain, 2018). Participating in such innovations allows the BOP communities to access means and opportunities that enhance their life standards and pursue entrepreneurship and sustainable development (Shahid et al., 2023).

The effectuation theory perspective can effectively explain the phenomenon of frugal innovation, which delineates the importance of using available resources instead of predictive strategies and planning (Sarasvathy, 2001). Operating with limited opportunities and unavailability of resources, frugal innovators exhibit adaptability and the ability to steadily develop their ideas to address the critical needs of the low-income context. This approach reflects the effectual mindset, where frugal innovators work within the boundaries of what is feasible, using ingenuity and existing means to create viable and scalable solutions.

Example: Tata Nano

In India, most households from middle-income families use two-wheeler bikes as their mode of transportation, which sometimes becomes highly unsafe, especially in adverse weather conditions. Owing a four-wheeler, such as a family car, seems to be a distant dream for several families hailing from the BOP segment. Recognizing this gap, Tata Motors, an Indian multinational company, aimed to create an affordable car for the masses, providing a safer and more comfortable alternative for middle and low-income groups. Through the process, material, and design innovation, Tata Motors, in the year 2008, developed 'The Tata Nano,' which cost about $2,000. Tata Nano was developed with only one vision—to create a safe, affordable car for India's emerging middle class (Anand, 2024). Furthermore, this frugal innovation was built with sustainability and fuel efficiency in mind, addressing environmental concerns as well. Tata addressed safety and environmental concerns by reengineering design and production processes to reduce costs, exemplifying sustainable practices in emerging markets (Radjou et al., 2012), and fulfilling the requirements of neglected groups while encouraging environmental responsibility.

Despite several constraints concerning high cost, availability of material, convenient design, integration of basic technology, and uncertainty of acceptance by the consumers, the development of Tata Nano reflects the logic of effectuation theory by using available resources and focusing on affordable, practical solutions for low-income families. This effectual mindset enabled the creation of a viable, scalable, and affordable car, demonstrating how effectuation helps navigate uncertainty and resource limitations through frugal innovations (Nielsen & Lassen, 2012).

7.2.5 *Jugaad Innovation*

Jugaad means an innovative solution or a workaround (Prabhu & Jain, 2015). It also asserts the art of improvising solutions with minimum resources (Ananthram & Chan, 2021). Jugaad innovation encompasses solutions with limited resources.

Though prior research claims that frugal innovation and jugaad pertain to analogous phenomena, some scholars have recognized jugaad as unique to India, possessing specific attributes such as frugality, flexibility, and inclusion (Radjou et al., 2012). Jugaad can make essential products and services for the masses by accentuating cost-effective and sustainable development. Further, jugaad can enhance the participation of the people by encouraging grassroots enterprise where indigenous knowledge can be employed to develop a sustainable economy. Jugaad innovations, due to their adaptability, can allow for-profit organizations to respond to critical societal problems and deliver value to BOP consumers.

Jugaad Innovation can be elucidated from the effectuation theory lens (Sarasvathy, 2001). In the context of jugaad innovation, the effectuation theory describes how entrepreneurs function in circumstances of uncertainty and scarcity of resources. Jugaad innovators bypass structured or predictable processes when bringing their innovations to life. With the resources, skills, and networks at their disposal, jugaad innovators ingeniously develop solutions that address the needs of underserved populations. Moreover, Jugaad innovations arise from iterative processes, resulting in cost-effective and practical solutions tailored for resource-limited settings (Shepherd et al., 2020). This adaptability is central to both theories, as effectuation emphasizes leveraging existing resources and responding flexibly to challenges, which mirrors the iterative and resourceful nature of Jugaad innovations. Consequently, effectuation offers a pertinent framework for comprehending the improvisational and adaptable characteristics of Jugaad innovation.

Example: 'Jugaadu Kamlesh'

In an episode of Shark Tank India, a renowned television program that offers a venue for entrepreneurs to showcase their concepts to prospective investors, an entrepreneur named Kamlesh Nanasaheb Ghumare, referred to as 'Jugaadu Kamlesh,' attracted considerable attention with the jugaad innovation designed to assist farmers in rural areas. Hailing from a modest village in Maharashtra, India, Kamlesh developed a low-cost pesticide spraying machine that alleviated the health-related difficulties encountered by small-scale farmers (Javaid, 2022). Kamlesh's example concords well with the effectuation theory, which emphasizes leveraging available resources to generate innovative solutions. With limited means, Kamlesh adapted local farming knowledge and improvised a cost-effective solution that directly addressed the critical community needs. Despite resource limitations, the ability to iterate and refine the product exemplifies Kamlesh's effectual approach to entrepreneurship, where innovation emerges from creativity, flexibility, and the intelligent use of existing resources (Sarasvathy, 2001).

While each type of innovation plays a significant role in empowering the BOP, they possess unique characteristics and are most effective in specific contexts. Understanding these distinctions can help stakeholders select the most appropriate innovation approach for addressing challenges within BOP communities. Table 7.1 clearly exhibits the types of innovations, their unique characteristics, and the optimal contexts in which they are most effective. In addition, Table 7.2 summarizes the type of innovations, connected theories, and specific examples.

The successful implementation and scaling of social, grassroots, reverse, frugal, and jugaad innovation relies on a well-structured and supportive ecosystem. Each innovation type depends on several factors to effectively empower the populace forming the BOP. The following section outlines the key elements constituting the support system in influencing the innovations responsible for meeting the needs of underserved communities.

7.3 Ecosystem Enablers Influencing Innovations in BOP

7.3.1 Government Policies and Regulation

Supportive government policies can greatly stimulate innovation. Governments can promote innovation by supporting firms, providing tax advantages, and implementing legislation that fosters research and development. Tax credits for R&D expenditures can reduce the cost of innovation for firms, making it more feasible to invest in new technologies tailored to BOP needs (Westmore, 2013). Streamlining regulatory compliance for startups and social companies can facilitate innovative activities aimed at social development (Mendoza-Silva, 2021). Public–private partnerships (PPPs) are more effective when both sectors can collaboratively improve each other for optimal outcomes. PPPs can consolidate resources, distribute risks, and integrate expertise to foster innovation in neglected markets (Van Ham & Koppenjan, 2001). Moreover, policies that allow firms to protect their intellectual property during open innovation might significantly advantage the BOP (Patanakul & Pinto, 2014). Finally, excessively rigid intellectual property regulations may inhibit innovation by restricting knowledge sharing. A more adaptable strategy can foster innovation while safeguarding competitive advantages (Gould & Gruben, 1996).

7.3.2 Infrastructure

Infrastructure development necessitates collaborative efforts among governments, the corporate sector, and local entities (Sivak et al., 2011). Enhancing fundamental infrastructure is essential for the implementation and acceptance of novel approaches. Infrastructural development initiatives, including roads, trains, power, and telecommunication fibers, facilitate access to various markets, hence promoting the scaling of innovations (Straub, 2011). Enhanced transportation networks diminish the expense and duration of transporting goods and services, facilitating the accessibility of innovative items to remote regions (Zeng et al., 2022). A deficiency in essential infrastructure might obstruct the supply and distribution of innovative products, separating BOP communities from market prospects (Fagbemi et al., 2022). Investments in digital infrastructure, such as broadband internet and mobile networks, can facilitate technological innovations and enhance their deployment (Salemink et al., 2017). The widespread adoption of mobile phones in developing nations has significantly influenced financial inclusion via mobile banking services (Suri, 2017). Ultimately, digital platforms facilitate access to education and healthcare services, addressing disparities resulting from geographical limitations.

7.3.3 Access to Finance

Availability of capital is a crucial driver for innovators and consumers in the BOP markets. As the BOP segment can provide enormous business opportunities simply because of its sheer volume, institutional investors interested in financial, social, and/or environmental returns can envisage investments in the BOP (Tasavori et al., 2016). Microfinance organizations must persist in providing capital to enterprises and consumers who face challenges in obtaining capital from traditional banks, as such initiatives significantly alleviate the burdens of lower-income groups in achieving their objectives (Vassallo et al., 2019). Crowdfunding can also help innovators from the BOP to source funds directly from the public, providing additional developmental opportunities. Lastly, the governments must continue to work on financial inclusion initiatives that expand credit and savings options for the BOP, supporting economic empowerment (Venkatesh et al., 2017).

7.3.4 Education and Skill Development

Investing in education augments human capital and can yield enduring impacts on economic development (Cobo, 2013). Initiatives directed toward individual skill enhancement are essential for innovation and adoption. In addition, leveraging technical skills can help individuals acquire the critical competencies to establish solution-centric innovations. Creating platforms that facilitate entrepreneurial initiatives can promote and assist local businesses, cultivating a culture of innovation. Lastly, awareness campaigns have substantial potential to educate communities regarding innovative solutions and their benefits, fostering community acceptance and utilization.

7.3.5 Social Network and Community Participation

Social networks and community involvement provide impetus for the effective execution of innovations. Community leaders must advocate and facilitate collaborative networks that disseminate information and resources among innovators, improve efficacy, and expedite learning. Comprehending cultural subtleties can enhance the probability of success, as solutions are more readily embraced when they correspond with local values and customs. Involving communities in the process of innovation guarantees that solutions are pertinent and genuinely meet the needs of the BOP (Venkatesh et al., 2017).

7.3.6 Technology Transfer and Adaptation

Technology transfer and adaptation mechanisms help replicate the success of effective innovations across broader contexts. Inter-organizational and inter-community communication can enhance the rate of innovation diffusion (Nilakanta & Scamell, 1990). Consequently, innovations must be implemented pertinently within specific settings, which involves modifying the product to comply with particular circumstances, cultural norms, and legal regulations. Collaborations between local and large corporations can enhance capacity on several fronts and ensure efficient technology transfer (Venkatesh et al., 2017). The following section discusses how integrating new-age technology can envisage the success of innovations hailing for, from, and at the BOP segments.

7.4 Innovation Types and Their Relative Dependence on the Supporting Ecosystem

Accompanied by marketplace evidence and literature support, this section discusses varying dependence levels of different innovation types on the enablers in the supporting ecosystem. Table 7.3 briefly summarizes the critical role of enablers on each innovation type, with the '+' symbol indicating the increasing levels of dependence.

7.4.1 Social Innovation (Relative Dependence: +++)

Social innovation relies heavily on a well-coordinated and supportive ecosystem involving government entities, non-governmental organizations (NGOs), community groups, and private enterprises. The high relative dependence of '+++' reflects this extensive dependence. Social innovations often address complex societal challenges that require effective multi-stakeholder collaboration. Government policies and regulations are essential in creating an enabling environment by providing funding, legal frameworks, and institutional support (Galego et al., 2022). In addition, active community engagement ensures that innovations are relevant, culturally appropriate, and sustainable (Cajaiba-Santana, 2014). Without significant contributions from these ecosystem actors, social innovations may struggle to operationalize and achieve scalability.

The Barefoot College in Rajasthan, India, is a prime example of social innovation empowering the BOP. Founded by Sanjit "Bunker" Roy in 1972, the college focuses on training illiterate and semi-literate rural women, often grandmothers, to become solar engineers, artisans, dentists, and doctors in their communities (Mininni, 2022). The success of Barefoot College relies on extensive collaboration with government agencies, international NGOs, and local communities. Government support has come in the form of policy endorsements and funding for rural electrification projects. International organizations like UNESCO and UNICEF have partnered with the college to expand its programs to other countries. Community participation is central, as the programs are designed to meet local needs and are delivered in culturally appropriate ways.

7.4.2 Grassroots Innovation (Relative Dependence: +++)

With a relative dependence rating of '+++', grassroots innovations depend heavily on community participation and social networks, as they are often initiated by community members themselves (Gupta, 2020). Institutional support from NGOs, government agencies, and academic institutions is critical in providing resources, training, and market linkages (Gupta, 2012). An enabling and protective policy environment, such as incorporating intellectual property rights, can facilitate grassroots innovations (Seyfang & Smith, 2007).

In March 2000, the Government of India established the National Innovation Foundation (NIF), which provides institutional support, including funding, technical assistance, and intellectual property protection to rural innovators who develop solutions tailored to local needs. This government-supported effort has expanded multiple grassroots ideas, improving community development and capacity building within the BOP (Gupta, 2020).

7.4.3 Reverse Innovation (Relative Dependence: ++)

Reflected by a '++' relative dependence rating, reverse innovation requires moderate support from the enablers in the ecosystem. While multinational corporations primarily drive reverse innovations, support from government and regulatory bodies is necessary to navigate different legal and regulatory environments between emerging and developed markets (Govindarajan & Ramamurti, 2011). Ecosystem actors can facilitate the adaptation and scaling of innovations across different contexts by providing insights into local market needs and helping to bridge cultural gaps. However, the innovation process is mainly driven by corporate entities, and extensive community involvement is less critical compared to social or grassroots innovations (Govindarajan & Trimble, 2012).

The collaboration between GE Healthcare and the Indian government to promote the portable ultrasound machine in rural areas exemplifies supportive policies enhancing reverse innovation (Govindarajan & Trimble, 2012). Government initiatives to improve rural healthcare infrastructure and provide regulatory approvals facilitated GE's introduction of cost-effective medical devices, benefiting BOP communities and eventually being adapted for developed markets.

7.4.4 Frugal Innovation (Relative Dependence: ++)

Frugal innovation, reflected by a '++' relative dependence, moderately relies on the supporting ecosystem. Innovators working on frugal solutions often face resource constraints and may require support in terms of infrastructure development and policy incentives to overcome these limitations. Assistance from ecosystem actors, such as governments and NGOs, can help frugal innovators reach BOP markets through distribution networks, financing options, and by creating awareness (Pisoni et al., 2018). While supportive policies and infrastructure can facilitate innovation, frugal innovators may also operate independently within existing constraints. Therefore, the role of ecosystem actors is important but not as critical as in social or grassroots innovations.

The Embrace Infant Warmer is a notable example of frugal innovation. Due to the absence of advanced incubators, several premature and low-birth-weight infants in developing nations are succumbing to mortality. Acknowledging this problem, a team of innovators from Stanford University developed a low-cost infant warmer that costs only a fraction of the expensive incubators. This innovation has since been leveraged by developing fruitful partnerships with NGOs, healthcare institutions, and government agencies (Morgaonkar et al., 2021).

7.4.5 Jugaad Innovation (Relative Dependence: +)

With a relative dependence rating of '+', jugaad innovators heavily rely on personal creativity, local knowledge, and available resources to develop practical solutions to daily problems faced by BOP communities (Radjou et al., 2012). Jugaad innovations often emerge outside formal institutions and may not rely significantly on government policies, institutional support, or infrastructure (Prabhu & Jain, 2015). While community support and informal networks can be beneficial, jugaad innovations are less dependent on formal ecosystem actors than other innovation types. Thus, the criticality of ecosystem support is relatively low for jugaad innovation, highlighting the importance of individual initiatives.

Remya Jose, a schoolgirl from Kerala, India, invented a pedal-powered washing machine made from recycled bicycle parts and a discarded washing machine drum. The device allows users to wash clothes by pedaling, providing a solution for people without access to electricity. Remya's innovation received recognition from the National Innovation Foundation, which helped showcase her invention on various platforms. However, the innovation was developed using locally available materials with minimal external support, exemplifying the essence of jugaad innovation (Ananthram & Chan, 2021).

7.5 The Way Forward—The Technology Way

Looking forward, corporations must address the demands of the bottom of the pyramid not merely as corporate social responsibility but also as a means to establish innovative business models that foster growth and sustainable development (Varadharajan et al., 2022). Organizations targeting the BOP groups can capitalize on numerous untapped economic opportunities (Gupta, 2020). Integration of innovation with advanced technologies such as artificial intelligence (AI), blockchain, and the Internet of Things (IoT) is expected to yield a beneficial transformation for the BOP community (Mariani et al., 2023). If realized, these technology breakthroughs could effectively solve issues in low-income communities through tailored products and services. For example, Blockchain can facilitate financial inclusion by providing secure, decentralized platforms that enable unbanked individuals to access credit and financial services, as seen with platforms like BanQu—a platform that enables individuals without formal identification or banking history to create a secure economic identity (Marchant, 2021). Alternative technologies, such as AI, can significantly empower the BOP segment. To illustrate, the Omdena Grassroots AI program demonstrates how AI may address particular challenges in disadvantaged communities (Younger, 2020). By bringing together local talent, domain experts, and global AI practitioners, Omdena aspires to enable communities to counter challenges associated with disaster management, healthcare facilities, and environmental concerns using new-age technology. This approach helps to identify contextually appropriate and feasible solutions as individuals from BOP are also involved in developing the necessary goods and services. Another illustration of the advantages of technology for the BOP through innovation is the 'Hello Tractor', which employs IoT technology to link tractor proprietors with smallholder farmers who need tractor services but cannot afford them (Broom, 2021). The Hello tractor sensors deliver real-time information regarding usage, maintenance requirements, and location, thus enhancing resource efficiency.

Despite the potential of new technology to empower BOP communities, numerous barriers hinder their application. For instance, deficiencies in digital literacy and financial limitations may lead to individuals being excluded from acquiring the skills necessary to utilize AI apps or blockchain platforms (OECD, 2019). Furthermore, infrastructural deficiencies, such as inconsistent electricity provision and inadequate

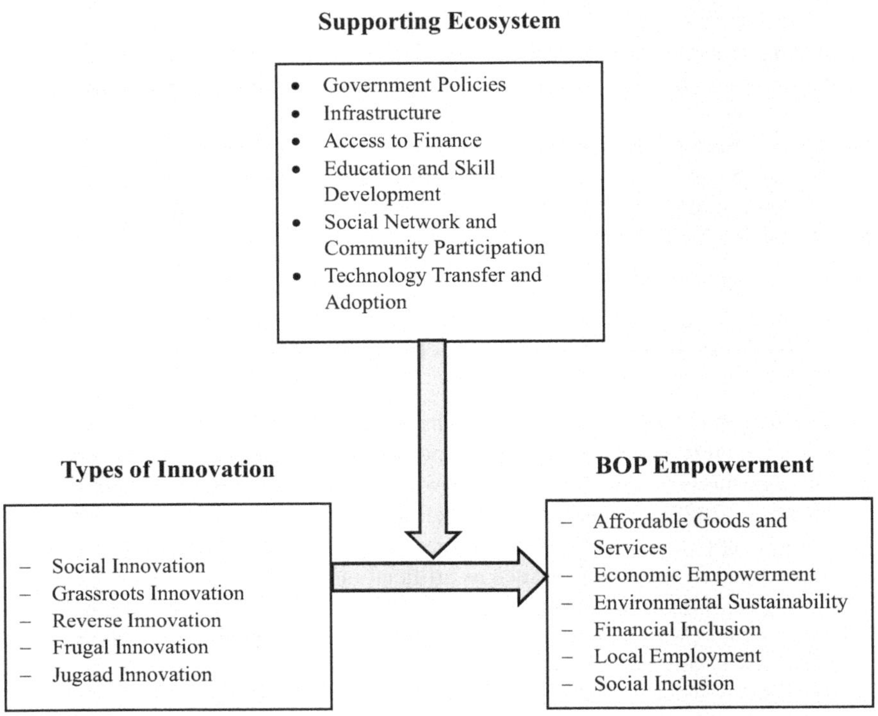

Fig. 7.1 Empowering BOP innovation and enabling ecosystem

internet access, provide considerable obstacles to implementing IoT devices and utilizing online services in distant or underserved regions (Sindakis & Showkat, 2024). Cultural and social barriers, including aversion to change, distrust of new technology, and linguistic disparities, can hinder acceptance and adoption, impeding the innovation process aimed at societal improvement for the general populace (McDermott & O'Dell, 2001). A coordinated partnership among governmental entities, private organizations, multinational enterprises, technological innovators, and BOP communities is essential to alleviate these obstacles.

Appendix

See Fig. 7.1 and Tables 7.1–7.3.

Table 7.1 Comparative analysis of innovation types

Innovation type	Unique characteristics	Most effective contexts
Social Innovation	• Collaborative efforts across sectors • Aims for social impact and systemic change • Community engagement	• Addressing complex social issues • Situations requiring multi-stakeholder collaboration • Enhancing social services
Grassroots Innovation	• Originates from local communities • Leverages indigenous knowledge • Focuses on local needs and solutions	• Rural or marginalized communities • Contexts where local participation is high • Preserving traditional knowledge
Reverse Innovation	• Innovations from emerging markets to developed markets • Cost-effective and simplified solutions	• Bridging technology gaps • Healthcare and technology sectors • Multinational corporations targeting new markets
Frugal Innovation	• Resource-efficient and cost-effective • Simplified design	• Resource-constrained environments • Providing affordable access to technology • Scaling essential services
Jugaad Innovation	• Improvisational and flexible • Individual ingenuity with minimal resources • Usually encompasses quick fixes to immediate problems	• Immediate and practical problem-solving • Environments with scarce resources • Informal sectors and daily life improvements

Table 7.2 Summary of innovations, connected theories, and examples

Type of innovation	Connected theory	Example	Brief description
Social Innovation	Resource-Based View (Barney, 1991)	Grameen Bank	Grameen Bank's microcredit model in Bangladesh leverages social trust and social capital to provide microloans without collateral, empowering the poor to engage in small businesses and improve their economic conditions
Grassroots Innovation	Community-Based Theory (Peredo & Chrisman, 2006)	Arunachalam Muruganantham's Low-Cost Sanitary Pads ("Padman")	Developed affordable and hygienic sanitary pads using local materials for rural women in India, addressing health issues and breaking social taboos around menstruation while also creating employment opportunities for women
Reverse Innovation	Disruptive Innovation Theory (Markides, 2006)	GE's Portable Ultrasound Machine	General Electric developed a low-cost, portable ultrasound machine for rural China and India, addressing the lack of affordable medical imaging. The device was later introduced in developed markets like the U.S. for emergency services and sports medicine
Frugal Innovation	Effectuation Theory (Sarasvathy, 2001)	Tata Nano—Affordable Car	Tata Motors created the Tata Nano, an affordable car designed for India's middle and lower-income groups, focusing on essential features, cost-effectiveness, and fuel efficiency without compromising safety standards
Jugaad Innovation	Effectuation Theory (Sarasvathy, 2001)	'Jugaadu Kamlesh'—Low-Cost Pesticide Sprayer	Kamlesh Ghumare developed an affordable, manual pesticide spraying machine for small-scale farmers in rural India, using available resources to address health and efficiency issues associated with traditional spraying methods

Table 7.3 The relative effect of supporting ecosystem on various innovation impact on BOP

Types of innovation	The relative of ecosystem actors	BOP impact areas
Social Innovation	+++	Access to finance, empowerment, and improved social services (Oeij et al., 2019)
Grassroots Innovation	+++	Community development, local solutions, and capacity building (Singh et al., 2021)
Reverse Innovation	++	Access to affordable products/services and improved healthcare (Govindarajan & Ramamurti, 2011)
Frugal Innovation	++	Affordable goods increased accessibility to technology (Weyrauch & Herstatt, 2016)
Jugaad Innovation	+	Practical solutions to daily problems enhanced livelihoods (Singh et al., 2012)

Legend +, ++, +++ indicate the increasing degree to which the supporting ecosystem actors play a critical role in various innovations impacting the BOP empowerment

References

Aksoy, L., Alkire, L., Choi, S., Kim, P. B., & Zhang, L. (2019). Social innovation in service: A conceptual framework and research agenda. *Journal of Service Management, 30*(3), 429–448. https://doi.org/10.1108/JOSM-11-2018-0376

Albert, M. (2019). Sustainable frugal innovation—The connection between frugal innovation and sustainability. *Journal of Cleaner Production, 237*, Article 117747. https://doi.org/10.1016/j.jclepro.2019.117747

Anand, A. (2024). From innovation to controversy: The untold story of the Tata Nano. https://medium.com/@abhisheksinghrana985/from-innovation-to-controversy-the-untold-story-of-the-tata-nano-bfde67ea1d57#:~:text=The%20Visionary%20Dream%3A%20Inception%20of,and%20commitment%20to%20uplift%20millions.

Ananthram, S., & Chan, C. (2021). Institutions and frugal innovation: The case of Jugaad. *Asia Pacific Journal of Management, 38*(3), 1031–1060. https://doi.org/10.1007/s10490-019-09700-1

Appadurai, A. J. P. C. (2000). Grassroots globalization and the research imagination. *Public Culture, 12*, 1–19.

Baishya, K., & Samalia, H. V. (2020). Extending unified theory of acceptance and use of technology with perceived monetary value for smartphone adoption at the bottom of the pyramid. *International Journal of Information Management, 51*,. https://doi.org/10.1016/j.ijinfomgt.2019.11.004

Barney, J. (1991). Firm resources and sustained competitive advantage. *Journal of Management, 17*(1), 99–120. https://doi.org/10.1177/014920639101700108

Berry, L. L., Danaher, T. S., Keiningham, T., Aksoy, L., & Andreassen, T. W. (2024). Social profit orientation: Lessons from organizations committed to building a better world. *Journal of Marketing*, 00222429241258495. https://doi.org/10.1177/00222429241258495

Broom, D. (2021). An app that allows farmers to rent tractors could revolutionize agriculture in Africa. Here's how. https://www.weforum.org/stories/2021/06/tractor-app-agriculture-farming-africa/

Cajaiba-Santana, G. (2014). Social innovation: Moving the field forward. A conceptual framework. *Technological Forecasting and Social Change, 82*, 42–51. https://doi.org/10.1016/j.techfore.2013.05.008

Cobo, C. (2013). Skills for innovation: Envisioning an education that prepares for the changing world. *Curriculum Journal, 24*(1), 67–85. https://doi.org/10.1080/09585176.2012.744330

D'Monte, L. (2016). How India powers GE's global innovation. https://www.livemint.com/Companies/H7FHZyvI5ix5E2V5nHpajN/How-India-powers-GEs-global-innovation.html

Fagbemi, F., Osinubi, T. T., & Adeosun, O. A. (2022). Enhancing sustainable infrastructure development: A boon to poverty reduction in Nigeria. *World Development Sustainability, 1*, Article 100006. https://doi.org/10.1016/j.wds.2022.100006

Forum, W. E. (2024). How microlending for women can address the gender gap and help alleviate global poverty. https://www.youtube.com/watch?v=tzNVf1zzl4M&t=116s

Galego, D., Moulaert, F., Brans, M., & Santinha, G. (2022). Social innovation & governance: A scoping review. *Innovation: The European Journal of Social Science Research, 35*(2), 265–290. https://doi.org/10.1080/13511610.2021.1879630

Gould, D. M., & Gruben, W. C. (1996). The role of intellectual property rights in economic growth. *Journal of Development Economics, 48*(2), 323–350. https://doi.org/10.1016/0304-3878(95)00039-9

Govindarajan, V., & Ramamurti, R. (2011). Reverse innovation, emerging markets, and global strategy. *Global Strategy Journal, 1*(3–4), 191–205. https://doi.org/10.1002/gsj.23

Govindarajan, V., & Trimble, C. (2012). Reverse innovation: A global growth strategy that could pre-empt disruption at home. *Strategy & Leadership, 40*(5), 5–11. https://doi.org/10.1108/10878571211257122

Gupta, A. K. (2006). From sink to source: The honey bee network documents indigenous knowledge and innovations in India. *Innovations: Technology, Governance, Globalization, 1*(3), 49–66. https://doi.org/10.1162/itgg.2006.1.3.49

Gupta, A. K. (2012). Innovations for the poor by the poor. *International Journal of Technological Learning, Innovation and Development, 5*(1–2), 28–39. https://doi.org/10.1504/ijtlid.2012.044875

Gupta, S. (2020). Understanding the feasibility and value of grassroots innovation. *Journal of the Academy of Marketing Science, 48*(5), 1552–7824. https://doi.org/10.1007/s11747-019-00639-9

Gupta, S., Kumar, V., & Karam, E. (2020). New-age technologies-driven social innovation: What, how, where, and why? *Industrial Marketing Management, 89*, 499–516. https://doi.org/10.1016/j.indmarman.2019.09.009

Hart, S. (2010). *Capitalism at the crossroads: Next generation business strategies for a post-crisis world*. Prentice Hall.

Hossain, M. (2018). Frugal innovation: A review and research agenda. *Journal of Cleaner Production, 182*, 926–936. https://doi.org/10.1016/j.jclepro.2018.02.091

Hossain, M., Simula, H., & Halme, M. (2016). Can frugal go global? Diffusion patterns of frugal innovations. *Technology in Society, 46*, 132–139. https://doi.org/10.1016/j.techsoc.2016.04.005

Javaid, A. (2022). Farmer by profession, jugaadu by choice: Meet Jugaadu Kamlesh, the shining star of Shark Tank India. https://www.jagranjosh.com/general-knowledge/jugaadu-kamlesh-shark-tank-india-1642779146-1

Kumar, H., & Bhaduri, S. (2014). Jugaad to grassroot innovations: Understanding the landscape of the informal sector innovations in India. *African Journal of Science, Technology, Innovation and Development, 6*(1), 13–22. https://doi.org/10.1080/20421338.2014.895481

Malodia, S., Gupta, S., & Jaiswal, A. K. (2020). Reverse innovation: A conceptual framework. *Journal of the Academy of Marketing Science, 48*(5), 1009–1029. https://doi.org/10.1007/s11747-019-00703-4

Mariani, M. M., Machado, I., & Nambisan, S. (2023). Types of innovation and artificial intelligence: A systematic quantitative literature review and research agenda. *Journal of Business Research, 155*, Article 113364. https://doi.org/10.1016/j.jbusres.2022.113364

Markides, C. (2006). Disruptive innovation: In need of better theory. *Journal of Product Innovation Management, 23*(1), 19–25. https://doi.org/10.1111/j.1540-5885.2005.00177.x
Marchant, N. (2021). This start-up is using blockchain to help smallholder farmers prosper. https://www.weforum.org/stories/2021/05/banqu-financial-inclusion-sustainability/
McDermott, R., & O'Dell, C. (2001). Overcoming cultural barriers to sharing knowledge. *Journal of Knowledge Management, 5*(1), 76–85. https://doi.org/10.1108/13673270110384428
Mendoza-Silva, A. (2021). Innovation capability: A sociometric approach. *Social Networks, 64*, 72–82. https://doi.org/10.1016/j.socnet.2020.08.004
Mininni, G. M. (2022). The Barefoot College 'eco-village' approach to women's entrepreneurship in energy. *Environmental Innovation and Societal Transitions, 42*, 112–123. https://doi.org/10.1016/j.eist.2021.12.002
Morgaonkar, V. A., Patel, D. V., Phatak, A. G., Nimbalkar, A. S., & Nimbalkar, S. M. (2021). Embrace versus cloth wrap in preventing neonatal hypothermia during transport: A randomized trial. *Journal of Perinatology, 41*(2), 330–338. https://doi.org/10.1038/s41372-020-0734-x
Nair, A. K., Tiwari, R., & Buse, S. (2017). Emerging patterns of grassroots innovations: Results of a conceptual study based on selected cases from India. In C. Herstatt & R. Tiwari (Eds.), *Lead market India: Key elements and corporate perspectives for frugal innovations* (pp. 65–95). Springer International Publishing.
National Innovation Foundation. https://nif.org.in/aboutnif
Nielsen, S. L., & Lassen, A. H. (2012). Identity in entrepreneurship effectuation theory: A supplementary framework. *International Entrepreneurship and Management Journal, 8*(3), 373–389. https://doi.org/10.1007/s11365-011-0180-5
Nilakanta, S., & Scamell, R. W. (1990). The effect of information sources and communication channels on the diffusion of innovation in a data base development. *Environment Management Science, 36*(1), 24–40. https://doi.org/10.1287/mnsc.36.1.24
OECD. (2019). *Measuring the digital transformation*. https://www.oecd-ilibrary.org/science-and-technology/measuring-the-digital-transformation_9789264311992-en
Oeij, P. R. A., van der Torre, W., Vaas, F., & Dhondt, S. (2019). Understanding social innovation as an innovation process: Applying the innovation journey model. *Journal of Business Research, 101*, 243–254. https://doi.org/10.1016/j.jbusres.2019.04.028
Patanakul, P., & Pinto, J. K. (2014). Examining the roles of government policy on innovation. *The Journal of High Technology Management Research, 25*(2), 97–107. https://doi.org/10.1016/j.hitech.2014.07.003
Peredo, A. M., & Chrisman, J. J. (2006). Toward a theory of community-based enterprise. *Academy of Management Review, 31*(2), 309–328. https://doi.org/10.5465/amr.2006.20208683
Pisoni, A., Michelini, L., & Martignoni, G. (2018). Frugal approach to innovation: State of the art and future perspectives. *Journal of Cleaner Production, 171*, 107–126. https://doi.org/10.1016/j.jclepro.2017.09.248
Prabhu, J., & Jain, S. (2015). Innovation and entrepreneurship in India: Understanding jugaad. *Asia Pacific Journal of Management, 32*(4), 843–868. https://doi.org/10.1007/s10490-015-9445-9
Prahalad, C. K. (2012). Bottom of the pyramid as a source of breakthrough innovations. *Journal of Product Innovation Management, 29*(1), 6–12. https://doi.org/10.1111/j.1540-5885.2011.00874.x
Purohit, S., Paul, J., & Mishra, R. (2021). Rethinking the bottom of the pyramid: Towards a new marketing mix. *Journal of Retailing and Consumer Services, 58*, Article 102275. https://doi.org/10.1016/j.jretconser.2020.102275
Radjou, N., Prabhu, J., & Ahuja, S. (2012). *Jugaad innovation: Think frugal, be flexible, generate breakthrough growth*. John Wiley & Sons.
Rahman, M. S., Mannan, M., & Amir, R. (2018). The rise of mobile internet: The adoption process at the bottom of the pyramid. *Digital Policy, Regulation and Governance, 20*(6), 582–599. https://doi.org/10.1108/DPRG-05-2018-0024

Raj, G., Feola, G., Hajer, M., & Runhaar, H. (2022). Power and empowerment of grassroots innovations for sustainability transitions: A review. *Environmental Innovation and Societal Transitions, 43*, 375–392. https://doi.org/10.1016/j.eist.2022.04.009

Salemink, K., Strijker, D., & Bosworth, G. (2017). Rural development in the digital age: A systematic literature review on unequal ICT availability, adoption, and use in rural areas. *Journal of Rural Studies, 54*, 360–371. https://doi.org/10.1016/j.jrurstud.2015.09.001

Sarasvathy, S. D. (2001). Causation and effectuation: Toward a theoretical shift from economic inevitability to entrepreneurial contingency. *Academy of Management Review, 26*(2), 243–263. https://doi.org/10.5465/amr.2001.4378020

Seyfang, G., & Smith, A. (2007). Grassroots innovations for sustainable development: Towards a new research and policy agenda. *Environmental Politics, 16*(4), 584–603. https://doi.org/10.1080/09644010701419121

Shahid, M. S., Hossain, M., Shahid, S., & Anwar, T. (2023). Frugal innovation as a source of sustainable entrepreneurship to tackle social and environmental challenges. *Journal of Cleaner Production, 406*, Article 137050. https://doi.org/10.1016/j.jclepro.2023.137050

Shepherd, D. A., Parida, V., & Wincent, J. (2020). The surprising duality of jugaad: Low firm growth and high inclusive growth. *Journal of Management Studies, 57*(1), 87–128. https://doi.org/10.1111/joms.12309

Sindakis, S., & Showkat, G. (2024). The digital revolution in India: Bridging the gap in rural technology adoption. *Journal of Innovation and Entrepreneurship, 13*(1), 29. https://doi.org/10.1186/s13731-024-00380-w

Singh, R., Gupta, V., & Mondal, A. (2012). Jugaad—From 'making do' and 'quick fix' to an innovative, sustainable and low-cost survival strategy at the bottom of the pyramid. *International Journal of Rural Management, 8*(1–2), 87–105. https://doi.org/10.1177/0973005212461995

Singh, S. H., Bhowmick, B., Eesley, D., & Sindhav, B. (2021). Grassroots innovation and entrepreneurial success: Is entrepreneurial orientation a missing link? *Technological Forecasting and Social Change, 164*, Article 119582. https://doi.org/10.1016/j.techfore.2019.02.002

Sivak, R., Caplanova, A., & Hudson, J. (2011). The impact of governance and infrastructure on innovation. *Post-Communist Economies, 23*(02), 203–217. https://doi.org/10.1080/14631377.2011.570050

Smith, A., Fressoli, M., & Thomas, H. (2014). Grassroots innovation movements: Challenges and contributions. *Journal of Cleaner Production, 63*, 114–124. https://doi.org/10.1016/j.jclepro.2012.12.025

Srivastava, A., Mukherjee, S., & Jebarajakirthy, C. (2020). Aspirational consumption at the bottom of pyramid: A review of literature and future research directions. *Journal of Business Research, 110*, 246–259. https://doi.org/10.1016/j.jbusres.2019.12.045

Straub, S. (2011). Infrastructure and development: A critical appraisal of the macro-level literature. *The Journal of Development Studies, 47*(5), 683–708. https://doi.org/10.1080/00220388.2010.509785

Suri, T. (2017). Mobile money. *Annual Review of Economics, 9*, 497–520. https://doi.org/10.1146/annurev-economics-063016-103638

Tasavori, M., Ghauri, P. N., & Zaefarian, R. (2016). Entering the base of the pyramid market in India. *International Marketing Review, 33*(4), 555–579. https://doi.org/10.1108/IMR-03-2014-0085

Tiwari, M. (2017). Exploring the role of the capability approach in social innovation. *Journal of Human Development and Capabilities, 18*(2), 181–196. https://doi.org/10.1080/19452829.2016.1271312

Van Ham, H., & Koppenjan, J. (2001). Building public-private partnerships: Assessing and managing risks in port development. *Public Management Review, 3*(4), 593–616. https://doi.org/10.1080/14616670110070622

Van Oost, E., Verhaegh, S., & Oudshoorn, N. (2009). From innovation community to community innovation: User-initiated innovation in wireless Leiden. *Science, Technology, & Human Values, 34*(2), 182–205. https://doi.org/10.1177/0162243907311556

Varadarajan, R., Welden, R. B., Arunachalam, S., Haenlein, M., & Gupta, S. (2022). Digital product innovations for the greater good and digital marketing innovations in communications and channels: Evolution, emerging issues, and future research directions. *International Journal of Research in Marketing, 39*(2), 482–501. https://doi.org/10.1016/j.ijresmar.2021.09.002
Varma, V. (2020). India's first solar-powered ferry Aditya wins global honour. https://indianexpress.com/article/india/indias-first-solar-powered-ferry-aditya-wins-gustave-trouve-award-6525729/
Vassallo, J. P., Prabhu, J. C., Banerjee, S., & Voola, R. (2019). The role of hybrid organizations in scaling social innovations in bottom-of-the-pyramid markets: Insights from microfinance in India. *Journal of Product Innovation Management, 36*(6), 744–763. https://doi.org/10.1111/jpim.12504
Venkatesh, V., Shaw, J. D., Sykes, T. A., Wamba, S. F., & Macharia, M. (2017). Networks technology, and entrepreneurship: A field quasi-experiment among women in Rural India. *Academy of Management Journal, 60*(5), 1709–1740. https://doi.org/10.5465/amj.2015.0849
Westmore, B. (2013). Innovation and growth: Considerations for public policy. *Review of Economics and Institutions, 4*(3), 50. https://doi.org/10.5202/rei.v4i3.128
Weyrauch, T., & Herstatt, C. (2016). What is frugal innovation? Three defining criteria. *Journal of Frugal Innovation, 2*(1), 1. https://doi.org/10.1186/s40669-016-0005-y
Younger, J. (2020). More ways for freelancers to prosper: AI freelancers at Omdena.com team up to solve tough social and economic challenges. https://www.forbes.com/sites/jonyounger/2020/08/14/more-ways-for-freelancers-to-prosper-ai-freelancers-at-omdenacom-team-up-to-solve-tough-social-and-economic-challenges/
Zeng, Z., Zhao, L., & Wang, X. (2022). Does improved transportation promote innovation? Evidence from China's cities. *Applied Economics, 54*(23), 2643–2657. https://doi.org/10.1080/00036846.2021.1992344
Zeschky, M., Widenmayer, B., & Gassmann, O. (2011). Frugal innovation in emerging markets. *Research-Technology Management, 54*(4), 38–45. https://doi.org/10.5437/08956308X5404007

Shaphali Gupta is a Professor of Marketing at Indian Institute of Management, Kozhikode. She is also a senior associate editor of 'The Marketing Strategy Journal,' and an associate editor of the 'Journal of Business Research' and 'Journal of Strategic Marketing'. She is also on the editorial board of 'The Journal of the Academy of Marketing Science' and 'Psychology & Marketing'. Dr Gupta's research interests are marketing strategy, Innovation for the greater good, Customer Relationship Management, new-age immersive technologies, customer-centricity, customer/stakeholder engagement, and sustainability. She has published her work in top-tier marketing journals (which includes AMA journals, UT Dallas, FT 50, ABS 4* and 4, and ABDC A*) such as the Journal of International Business Studies, Journal of the Academy of Marketing Science, International Journal of Research in Marketing (IJRM), Journal of Retailing, Journal of Public Policy in Marketing, Journal of Advertising, Industrial Marketing Management, Psychology & Marketing, Journal of Interactive Marketing, Journal of International Marketing and Journal of Business Research. She has also published managerial articles and cases in reputed magazines such as Harvard Business Review, AIB Insights, and Ivey Publishing. Before entering academia, she worked in the industry, and now she conducts management development programs for senior executives of major Indian companies regularly.

Rushabh Trivedi is a Post-Doctoral Fellow at the MICA. Formerly, he served as an assistant professor at the School of Business,Woxsen University, India. Dr Rushabh completed his PhD from the esteemed University of Hyderabad. He is also an Indian Institute of Management – Kozhikode Faculty Development Program alumnus. Before joining Ph.D., he completed his Master's in Business Administration (MBA) and Post Graduate Program in Management (PGPM) from ICFAI Foundation of Higher Education. Rushabh's research focuses on social entrepreneurship and workplace cognitions, examining mind wandering and daydreaming. He has a passion for research and has been invited as a resource person for faculty development programs, workshops, seminars, and webinars at esteemed universities. He has authored several research articles in international journals indexed in ABDC, Scopus, and Web of Science. He has also reviewed research papers for esteemed journals such as Career Development International, Journal of Creative Communications, and International Journal of Indian Culture and Business Management.

Chapter 8
Sustainable Luxury Brands Strategies for Base of the Pyramid Wellbeing

Amrita Bansal and Varsha Jain

Abstract Luxury brands maintain their tremendous market share by producing high-quality products and keeping an aspirational status for non-affluent customers. However, changing social and economic distrust have led to increased need in inclusivity of marginalized sections of the society. Thus, the chapter identifies and explores various innovative strategies implemented by the luxury/premium brands for the BOP communities' well-being. These strategies transcend the brand-consumer transactional relationships. BOP wellbeing is catered by inclusion of these communities as suppliers and employees to create value for the end consumers, which directly impacts the daily lives of BOP communities. As BOP individuals are involved in the brand's value creation process, hence, the strategies can go beyond the promotional tactics of the brands. Four different luxury brands are analysed to understand their strategies and tactics for the BOP wellbeing and inclusion. The case analysis focused on the similarities in the strategic level of the brands. FabIndia, Jaipur Rugs, Farm2kitchen, and Sabyasachi are the brands analysed for this chapter. The secondary data collected was integrated with the theories of PERMA (Positive Emotions, Engagement, Relationships, Meaning, and Accomplishment) framework of wellbeing (Ibrahim et al., 2023) and value chain business model (Porter, 1985). The integration of theories with brands' data helped the research to suggest five distinct strategies for BOP wellbeing. After analysing the secondary data through thematic analysis, the chapter proposes the themes: under Economic Empowerment category the themes are accessibility: building the baseline and blend of cooperative and corporate through collaboration, under Social Development Accessibility category the themes are educating consumers through stories to increase demand, end-to-end communication among stakeholders and lastly, under Skill Development and Sustenance category the theme is preservation and passing on the skills and cultural heritage. The tactics the brands apply are contextual and per the needs of the individuals at the base of the pyramid.

A. Bansal (✉)
IMI, New Delhi, India
e-mail: amrita.bansal@imi.edu

V. Jain
MICA, Ahmedabad, India
e-mail: varsha.jain@micamail.in

R. Singh and T. A. Wani (eds.), *Marketing to the Base of the Pyramid*,
https://doi.org/10.1007/978-981-96-7105-2_8

Keywords Luxury brands · Base of the pyramid (BOP) · Value chain · Framework of wellbeing · Inclusive business models

8.1 Introduction

Traditionally, luxury brands focused majorly on the exclusivity of their offerings, high prices, and catering to affluent consumers. These exclusive brands maintained an excellent global market share by producing high-quality products and kept an aspirational status for non-affluent customers (Ko et al., 2017). However, scholars have pointed out that there is a growing interest in inclusivity and marginalized sections of the society by media and current generations.

The expectations of the publics from luxury brands have evolved. Mindful consumption is an important aspect of their consumer journey (Manchanda et al., 2023). Hence, they expect the brand they associate with to be inclusive and mindful of the society in which the brands operate. These expectations are beyond the transactional value of product offerings and status aspirations. It is the brands to positively impact society and the environment (Hanson et al., 2018). A brand that fails to meet these expectations of the current consumers is alienated by these segments, which leads to a loss of market share and competitive edge in this conscious and mindful market (Nadanyiova & Das, 2020). Hence, brands like Armani, Chanel, Gucci and more have started working on inclusive business models and sustainability (Staff, 2021).

Further, previous literature also looks at luxury brands as societal trend setters as other companies copy their innovations and processes (Kunz et al., 2020). Thus, it has been emphasized by scholars' luxury brands' imperative role in inclusive production processes (Joy et al., 2012). United Nations' Sustainable Development Goals (SDG) provide abundant preference for inclusivity and wellbeing (Filho et al., 2017). SDG goal eight directly aims at inclusivity of all different sections of the society. It directly states, "*Foster sustained, inclusive and sustainable economic growth, full and productive employment and decent work for all*" (Goal 8 | Department of Economic and Social Affairs, n.d.). The idea is to include different parts of society, especially the BOP (Base of the Pyramid), for the overall development of the world (Gupta & Kanungo, 2022). Moreover, luxury brands, with their immense resources, can play a vital role in this inclusive development (Lee et al., 2024). Thus, brands like Hermes have also started changing their business strategies to procure alternative and sustainable raw materials and localized supply chain for sustainable value chain (Birch, 2024).

Hence, the change in consumer and economic expectations has transformed luxury brands' objectives and strategies (Atkinson & Kang, 2021). From fashion luxury brands to high-end automobile brands, all are rapidly finding more inclusive ways to operate their businesses (Debenedetti et al., 2024). One critical initiative is including BOP (Base of the Pyramid) individuals in the value chain of luxury brands. Including

BOP is not restricted to moral obligation but provides a strategic and tactical advantage to luxury brands (Bhattacharya & Dalal, 2024). The advantage originates from the participation of the BOP in the value chain and bringing in their unique talent for the brands as a point of differentiation (Lappeman et al., 2019). Additionally, it enables the brands to contribute to the wellbeing of the marginalized section of society (Cristini et al., 2022).

However, the integration of BOP with luxury brands is arduous. Luxury brands' orientation is to be exclusive, have high prices, and cater to a specific, small, affluent section of society (Ko et al., 2017). Integrating the BOP into the value chain of luxury brands requires rethinking the traditional business models. In place of including BOP as a consumer segment for the brands, the managers need to consider BOP as the value creation partner in the value chain of the businesses (Schoneveld, 2020). Rather than looking at them as consumers, the BOP can be in the roles of suppliers or employees. This inclusion in the value creation process will help the BOP to elevate their status and self-reliance (Reficco & Márquez, 2009). Luxury brands will have a more sustainable image in the minds of their consumers and can also gain a competitive edge in the market.

To do so, luxury brands can apply different strategies to become inclusive for BOP wellbeing. The previous studies discuss the perceived social value of the consumers with respect to luxury brands, impact of social messaging by luxury brands (Kang & Sung, 2021) and luxury brands' CSR-related research (Olšanová et al., 2021). There is a dearth of research with respect to luxury brands and BOP inclusion for their wellbeing. Hence, this chapter will identify and explain in detail the strategy luxury brands can use to include BOP in their value chain phenomenon, which leads to BOP wellbeing. The strategies will help the managers of luxury brands to create processes to include BOP artisans and farmers with distinct skills to participate in the value chain. The inclusion will lead to BOP wellbeing and brands will get exclusive access to their craftsmanship. The following section will discuss and present arguments concerning BOP wellbeing, luxury brands' inclusivity to identify the literature gap and explain the objective.

8.2 Luxury Brands and BOP Inclusion

Luxury brands are majorly described as the market offerings of high quality, providing emotional or/ utility benefits to the consumers, connected strongly with prestige by offering high-quality craftsmanship or service quality, priced at a premium level, and aspirational (Keller, 2017). However, the new luxury concept explains luxury goods and services as brands with higher quality and aspirations than the other players in the category within the reach of consumers and not exuberantly expensive (Husain et al., 2021). Luxury consumption in the current scenario has changed because of the changing aspirations and realities of the consumer segments (Kumar et al., 2022).

Hence, in the current scenario, luxury consumption is about consumer and other stakeholders' participation and multiple engagement channels (Rosenbaum et al., 2019). Moreover, the consumers' expectations of luxury brands have been exceeded. They demand brands not only to provide quality products and prestige to their consumers but also to become champion corporate citizens of society (Atkinson & Kang, 2021). Hence, the luxury brands have started focusing on the BOP section of the society as producers for inclusivity and positive social impact (London et al., 2009).

Brands delve into producer-oriented BOP interventions where the BOP producers are included to create new products or services for the end consumers (London, 2008). This kind of inclusion helps the BOP to enhance their current techniques, learn new skills and gain a better standard of living (Porter & Kramer, 2007). Additionally, these interventions from luxury brands help the BOP producers to navigate through factor markets (Reyes-Menendez et al., 2021). Luxury brands either provide raw materials on their own or resources to buy the raw materials to produce finished goods by BOP, which helps in navigating the factor markets easily (Haworth & Hart, 2007).

By providing equitable opportunities and compensation to BOP producers, brands try to make more inclusion and participation (Moore et al., 2022). It helps the BOP producers to understand the current market trends and consumers' expectations as they work with the organizations who have access to the knowledge and information (Vishwanathan et al., 2009). Luxury brands create a bridge between the informal context of the BOP producers' skills and practices and formal organizational and market set up (London et al., 2009).

Still, luxury brands face various challenges to be seen as inclusive. The major challenge is creating an inclusive perception of luxury and BOP producers. Most literature still argues that luxury is more exclusive than inclusive. Apart from the above mentioned processes, there is a long way to go. There is a need to create strategies that make BOP producers part of the value chain of luxury brands which leads to their wellbeing.

8.3 BOP (Base of the Pyramid) Producer Segment and Inclusive Wellbeing

The four billion individuals' base segment of the pyramid is critical for modern-day brands and social sciences experts (Chmielewski et al., 2018). BOP is a critical mass because of its potential as a vast market and fundamental requirement to work for their well being as an inclusive society (Follman, 2012). The relationship between brands' goals and BOP well-being is complex. This complexity confronts marketers and policymakers to reevaluate their traditional methods and business models (Sinha et al., 2017).

Earlier, the literature around BOP actively engaged in the argument of considering the individuals at the base of the pyramid as the untapped market and how the brands can target them (Prahalad, 2011). Hence, BOP was seen as having an untapped consumer-centric role. However, this role has progressed in the last decade to producers and entrepreneurs (Kumar et al., 2022).

The BOP segment is identified by two main features: low income and limited availability and accessibility of resources. The literature has pointed out that non-inclusive business models cannot solve these problems of BOP (Ghosh & Rajan, 2019). These models are formed from the orientation of the ability of the segments to pay for the quality goods and services provided by the brands. Hence, there is an urgent need for inclusive and innovative business models that can answer the unique needs of BOP (Rosenstock et al., 2020).

Development-oriented literature explains inclusive business models as strategies and processes to include low-income groups or BOP as providers of profitable markets, potential workforce, and nourishers to the supply chain of the businesses (Derks et al., 2022). Hence, the inclusive business models look at BOP beyond consumers. The individuals at BOP can also participate in the business processes as producers, distributors, suppliers, and skilled labour, which further can help in the socioeconomic development of these sections of society. Furthermore, it has been argued that real inclusion means finding dignity, credibility, and equity for the BOP while working in these business models. Additionally, practitioners understand that inclusive business models align with businesses' core values and goals (Kolk et al., 2013). The alignment between the business goals and BOP development is gained through contextual relevance, innovative tactics, and scalable programs, resulting in BOP wellbeing (Shukla et al., 2023).

An individual's wellbeing stems from the ideology that can explain what is suitable for a person. This notion is further advanced by adding optimal functioning to the good (Rahman et al., 2013). The wellbeing of an individual is the combination of mental, emotional, and physical health. Additionally, in attaining a good quality of life, there is a need for self-respect and dignity (Steptoe et al., 2014). The entrepreneurship success of the producers helps them to improve their economic conditions, which also improves their quality of life (Kumar et al., 2021). The collaboration between the BOP producers and brands, leads to engaging activities with high rewards for the BOP producers. This enhances the wellbeing of the BOP producers. Further, the established literature also argues that the progress in wellbeing is the strong signalling of development (Viswanathan et al., 2018). Hence, an inclusive value chain of businesses can help in enhancing the wellbeing of the BOP segment.

8.4 Value Chain and Inclusivity

The value chain is where various interconnections between brands and their partners create consumer-centric products through various functions (Lashitew et al., 2018). The value chain and its sustainable development have evolved tremendously in the last few years. More interventions are being done to strengthen the nodes between the agents and the brands to get better output. Further, more businesses are trying to place marginal sections of society like BOP, women, farmers, and more as their agents (Agyekumhene et al., 2020). This inclusivity in the value chain demands businesses to generate solutions with diverse stakeholders and develop mutually beneficial strategies to better the performance of the value chain. The literature argues that by working closely with companies, the marginalized stakeholders can enhance their performance and sustain their livelihoods even after the end of a particular intervention (Nielsen & Aagaard, 2021).

Inclusivity in the value chain, especially for luxury brands, must be understood from the viewpoint of integrity (Karaosman et al., 2018). Including BOP in the value chain of the business requires ethical practices with a combination of transparency and accountability in all the various processes of business functions (Jaén et al., 2020). The inclusive value chain helps the vulnerable section strengthen its linkages with the businesses, increase income, and gain hands-on experience with quality processes and services (Roberson & Perry, 2021).

However, the biggest challenge for businesses to be inclusive with their value chain is the subjectivity related to BOP issues and needs. It is evident that BOPs in different parts of the world have unique problems and requirements; hence, businesses cannot have standardized programs to solve BOP issues. There is a need for a deeper understanding of the local actors and circumstances, making the situation highly contextual (Ranjan et al., 2021). Further, there is a limited understanding of the capacity of BOP to continue working with businesses in a market-oriented environment for the long term. Apart from the capacity, the risk-taking capability of the BOP also needs to be analysed and understood. Most importantly, because of the high subjectivity of the inclusive value chain, it is imperative to have a constant dialogue with the actors of the value chain (Dembek et al., 2019).

These challenges with the inclusive value chain for BOP make it unique and need a more robust, deep understanding of both sides working together. Business orientation is different from BOP's orientation and expectations, especially for luxury brands. To become sustainable and work for a positive social impact, luxury brands are trying to be inclusive. However, the highly subjective nature of the BOP population in different parts of the world requires more work and deeper understanding. Thus, this chapter will explore various strategies luxury brands can apply to be inclusive and work better with BOP for their wellbeing and positive social impact.

These strategies are formed after performing qualitative case analysis (Alam, 2020) on various luxury and premium brands operating in India. As per the case analysis, the chapter focuses more on the similarities in the strategic level of these brands and the differentiations around their tactical steps. FabIndia, Jaipur Rugs, Farm2kitchen, and Sabyasachi are the brands analysed for this chapter. A brief background on the four brands is presented in Table 8.1.

Further, the research analysed various theoretical frameworks of inclusive value chain and wellbeing to create a strategic impact. After analysing multiple theoretical frameworks, the chapter drove its understanding of the luxury brand strategies for an inclusive value chain for BOP wellbeing from PERMA (Positive Emotions,

Table 8.1 Background of luxury/premium brands studied and analysed for the research

Luxury/Premium Brands	Background
FabIndia	The brand is one of the leading retail companies known for its handloom fabrics and traditional Indian clothing line. The brand is not limited to clothing; it also produces and distributes home goods, organic foods, and personal care products. A unique selling point (USP) is to blend modern style with traditional designs and materials. The brand has an inclusive business model where the rural artisans and craftsmen are part of the company's value chain
Jaipur Rugs	Jaipur Rugs is a luxury brand that produces the best handmade rugs for affluent consumers. The majority of the consumers are based out of the USA and UK. They have a strong online and offline presence, showcasing their high-end products. To preserve the art of hand-knitted rugs by the rural women artisans, the company integrated traditional craftsmanship with modern business practices and channellized their resources for the upliftment of women artisans
Farm2kitchen	A brand that started in one city in India is now an international phenomenon. The core of the business is to provide fresh and high-quality produce to affluent consumers by connecting them directly to the farmers. The company's vision is to close the gap between nutrition and agriculture. The brand mainly includes small and marginal farmers from BOP in their value chain
Sabyasachi	One of India's most celebrated fashion designers, known for its luxurious designs, especially bridal and occasions, is the Sabyasachi brand, which works at the intersection of Indian craftsmanship and contemporary. The brand is known for its opulence design, elaborate embroidery, and in-depth understanding and appreciation for India's diverse textile heritage

Engagement, Relationships, Meaning, and Accomplishment) framework of wellbeing (Ibrahim et al., 2023) and value chain model by Porter (1985), also its industry 4.0 version (Kohnova & Salajova, 2023). PERMA explains that a person's definition of a good life varies; hence, it is highly subjective. Additionally, the framework helps enable individuals to choose what is good for them rather than making a choice for themselves. PERMA and its elements help make an informed choice through positive emotions, engagement with various activities, fulfilling relationships, meaningful purpose, and accomplishments. The value chain is a strategic tool for businesses to plan and analyse multiple activities to accomplish their purpose. These activities are further divided into primary and secondary activities. Primary activities directly affect the creation of products or services, the consumers' buying process, and the after-sale activities. Secondary activities are the support system for the primary activities, which provide added value to the market offering. These activities include procuring raw materials, various business infrastructure, technology, and human resources throughout the value chain. The combination of these two frameworks helped in coming up with strategies that are inclusive and work for the BOP's wellbeing. The marriage of these two frameworks defined the activities performed

Table 8.2 Thematic analysis table

Category	First order theme	Second order theme
Economic Empowerment	*Accessibility: Building the Baseline*	Market accessibility, financial independence
	The blend of cooperative and corporate through collaboration	Structural and operational integration, Direct supplier connections, ERP
Social Development Accessibility	*Educating Consumers Through Stories to Increase Demand*	Brand facilitated processes for deep relationships among stakeholders, consumer education
	End-to-End Communication Among Stakeholders	Trust building, Sense of achievement
Skill Development and Sustenance	*Preservation and passing on the Skills and Cultural Heritage*	Training and skill development of future generations, sense of pride

Source Developed by Authors

by the luxury brands and BOP together to attain profitability for the brands and the wellbeing of BOP.

The data was collected from the secondary sources of the brands mentioned in Table 8.1. The secondary sources were company reports, news articles, case studies, company websites, social media reports of the brands (Ruggiano & Perry, 2019). All the sources used were from the year 2015–2024. This timeline was maintained to understand the current and sustained trends of the luxury brands. Further, a detailed thematic analysis was performed. The first order of themes was culled out with the help of framework of wellbeing (Ibrahim et al., 2023) and value chain model by Porter (1985), also its industry 4.0 version (Kohnová & Salajová, 2023). Further, second order themes were developed with more granular analysis of the secondary qualitative data through an iterative process. This helped us to identify different strategies implied by the brands in various contexts and the basis of those activities (Maxwell, 2005). The analysis helped us in identifying the common themes of the data which are presented in the following section (Rishi et al., 2015). A brief glimpse of the results from the thematic analysis is presented in Table 8.2.

The following section will explain these strategies in detail.

8.4.1 Accessibility: Building the Baseline

One of the significant issues for the BOP is the lack of access to the elements needed for their development and wellbeing. The accessibility of various markets, which are not restricted by geography, provides immense opportunities for the BOP. Because of their vast retail chains and online and offline presence, the luxury brands should be able to provide the products created by BOP in various markets. FabIndia provided its artisans access to multiple markets by adding the products they developed into their retail chains. Additionally, the brand encouraged artisans to diversify their products to incorporate modern style with their traditional craftsmanship. This strategy also helps gain market knowledge and produce more consumer-centric products. Likewise, Farm2Kitchen also provides insights regarding market demand mechanisms, assisting farmers to plan their crops and production cycles more effectively.

As accessibility to markets is essential, providing BOP with new knowledge and training is crucial. The latest knowledge provides hopefulness to the BOP, which is a vital element of positive emotions within the wellbeing of individuals. Luxury fashion brands like Sabyasachi conduct training programs for artisans to understand the new trends and techniques of the fashion world and, help the artisans to make their products more marketable. Jaipur Rugs trains the artisans in other activities, like how to use technology for better processing; this kind of training takes artisans' knowledge beyond rug-making craftsmanship. Also, Farm2Kitchen provides farmers with training on sustainable agricultural practices. The training improves their productivity and the quality of the produce. These training programs help individuals sustain growth, which is good for the brand and the wellbeing of the BOP.

Financial independence is necessary for any individual's wellbeing. If the growth depends on other entities, the individual is not empowered. Hence, for BOP, it is crucial to be financially independent. Farm2Kitchen provides financial aid in the form of micro-loans to the farmers. These loans help the farmers make their own decisions about investment as they understand their environment better. Jaipur Rugs firmly believes in lending money to contractors who demonstrate excellence in their jobs and a positive attitude, enabling them to play a crucial role in manufacturing. Many contractors experience increased wages, social positions, and skills because they recognize their talents. All the other three brands pay timely incentives to the artisans. Due to intermediaries, the payments were seasonal and not per the industry standards. However, paying the artisans monthly as their own employees provides artisans more freedom and space to invest for a better future.

To summarize, luxury brands need to provide accessibility to new markets, knowledge, and finances to the BOP for better inclusion into the value chain of their businesses. This accessibility provides better quality products to the brands, which becomes a differentiated factor for them in the market. For BOP, it offers more freedom to make decisions for their wellbeing.

8.4.2 The Blend of Cooperative and Corporate Through Collaboration

Combining cooperative and corporate features is necessary for luxury brands to include BOP into their value chain efficiently. As mentioned above, luxury brands provide the BOP with access to markets, knowledge, and funds, one of the most crucial features of a genuine cooperative organization. However, accessibility alone is unable to accomplish wellbeing. Brands and BOP need to engage and collaborate for a positive social impact. The brands must equate the BOP with their corporate processes and other stakeholders.

FabIndia has established Supplier Region Companies (SRC) to work directly with the artisan suppliers. The SRCs help the brand to have a direct connection with the artisans, which allows the processes to be transparent and efficient. Additionally, the SRC makes sure that the artisans are treated fairly.

Jaipur Rugs has created a business model that increases the collaboration with the artisans and helps both parties attain various benefits. The brand has gained flexibility by turning most of its processes into decentralized techniques. The model' pay for performance' is applied to the production processes of the rugs, where the producer determines the quality and volume of the final output. The company does not own the production assets and machinery but the producers, which helps the brand to be less capital intensive and artisans to own their assets. Enterprise Resource Planning (ERP) connects the brand with more than 40,000 artisans in different parts of the country. The connectivity helps the artisans to be part of the company's IT infrastructure, which provides them more space to interact with the brand and allows the brand to run its business uninterrupted.

Farm2Kitchen uses a digital platform to connect farmers with consumers. This platform includes online ordering, payment processing, and logistics management features. Since many BOP farmers have access to mobile phones, the platform is mobile-friendly, enabling farmers to manage their sales and track orders easily.

To summarize, the brands need to do technical and strategic interventions to equate BOP with their processes. The engagement helps the BOP individuals deploy their skills and strength for the complex tasks, leading to wellbeing by completing tasks. Brands can also collaborate more profitably with these individuals as they get efficient output.

8.4.3 End-To-End Communication Among Stakeholders

Relationships are a critical aspect of an individual's wellbeing. Relationships help individuals experience various emotions in their lives. It also leads to belongingness and the feeling that being part of a group is an accomplishment. Hence, to initiate such deep relationships, luxury brands create processes to communicate better with BOP. This communication is not limited to the company; it includes other suppliers, employees of different departments, and consumers. The brands perform various activities to orchestrate this end-to-end communication for deeper relationships.

All four brands promote the culture in their organizations to establish a strong line of communication with BOP to have deeper relationships. These activities help employees of different departments connect with farmers, artisans, and other stakeholders. Other kinds of activities also connect the artisans and farmers of BOP to communicate with the end consumers.

Jaipur Rugs organizes fests or 'artisans mela' (carnival) at different locations where artisans meet employees of various company departments. These meetings help the artisans understand the employees' contributions and roles in the rug-making and selling business. Understanding the employee's contribution helps the artisans to have a feeling of belongingness. And the employees of different departments appreciate their contribution to the lives of the artisans, which gives them a sense of accomplishment. Jaipur Rugs also shares the stories of the artisans and their journey of making a rug with the consumers. These stories are narrated by the artisans, who also share their experiences and aspirations behind making the rugs. The stories are shared with the consumers once they buy any of these products, and they are encouraged to provide their own experience with the rugs, which are also shared with the artisans. This shared experience helps the consumers have a more fulfilling experience with the purchase, and the artisans feel a sense of accomplishment to see the products they make appreciated by the consumers.

Similarly, Farm2kitchen also connects farmers directly with the consumers, firstly to eliminate the mediator to get a better share of the earnings and secondly to have a better relationship between the farmers and the consumers. The quality of the crops and the work of the farmers are shared with the consumers on the app and other social media handles of the brand. This information helps consumers trust the quality of the produce and understand the importance of farmers up and close. The farmers get a better share of the revenue and a sense of belongingness with this share of information.

Under the Sabyasachi brand, the designers and other teams also interact with the artisans directly to exchange their expertise with each other. To summarize, to create deeper relationships for the wellbeing of the BOP, the brands must create various avenues for communication among the employees, suppliers, and consumers. All the stakeholders better understand and feel a sense of achievement and belongingness after the various interactions facilitated by the luxury brands.

8.4.4 Educating Consumers Through Stories to Increase Demand

Consumers are the most critical participants in the value chain of the business. The base of the business process is to serve consumers' needs and exchange value with them. For luxury brands, the value exchange with the consumers is beyond the utility of the product/service. The products' quality and the brand's exclusivity are the elements demanded by the consumers from luxury brands. To maintain its exclusivity, the brands need to differentiate themselves uniquely from the competitors. Thus, luxury brands create positive social impact to differentiate themselves and be more sustainable. The positive impact is made by including BOP in the value chain. Consumers need to be educated to feel better about their ownership of luxury brands.

Luxury brands communicate with consumers effectively to make them understand their initiatives. This kind of communication mostly happens through storytelling. The brands create various narratives about the BOP and their participation in creating consumer value. The stories of the BOP individuals create a positive impact for the consumers and help them to choose the brands they are motivated to select. These stories of the products made by skillful individuals lead to increased consumer demand as they feel satisfied with their brand decisions.

Farm2Kitchen runs various campaigns to educate consumers about the benefits of buying directly from smallholder farmers, including their purchasing decisions' social and environmental impact. The business strongly emphasizes supply chain transparency so that customers understand their products' specific origins and production processes. Most of these campaigns have stories of the farmers and how they produce the product from start to end. These stories are personal narratives that impact the consumers better than brand-standardized content.

Jaipur Rugs also includes the narratives of the artisans and their personal stories with each rug they sold. The consumers treat these stories as part of their brand experience. Once the consumers place rugs in their homes, they become part of their conversations and stories with people inquiring about the carpet. Consumers feel more connected to the brand, and it generates positive word of mouth offline and online and increases the chances of consumers becoming brand advocates. FabIndia also promotes the stories of the artisans in their campaigns to position themselves better among their competitors. The artisans' narratives lead to more demand from the consumers.

To summarize, luxury brands educate consumers via the stories of farmers' and artisans' journeys of creating products. These stories provide more value to the consumers apart from the product purchase. This information and understanding of product creation processes helps the consumers understand and appreciate the brand's work towards the positive societal impact and BOP wellbeing. This understanding leads to more demand for advocacy from consumers.

8.4.5 Preservation and Passing on the Skills and Cultural Heritage

The strategies formulated by the luxury brands for BOP inclusion should go beyond the present participation of the individuals. The brands work for the skill development of the farmers and artisans; however, these measures are still beneficial for both the suppliers and the brands in the short term. After analysing all four brands, it is preferred that the brands also take measures to develop the skill set for future generations.

Due to a lack of opportunities, the present generations of artisans and farmers do not teach and motivate their future generations with the unique skills they have expertise in. They want future generations to learn other skill sets so they do not suffer like the others. Due to this uncertainty in income, many unique arts and skills are fading away as the newer generations have not learned them from their parents.

Hence, the brands must have a long-term plan for the long-term sustainability of the craftsmanship and brands' unique differentiation.

Farm2Kitchen organizes training programs open to other family members of the current farmers they do business with. It helps the company get more people to connect with their business, and the future generations feel more confident with the organized training systems. Further, Jaipur Rugs also offers various programs for the health and education of the artisans' children. Additionally, when the artisans visit the company's office, they are encouraged to bring their children, which helps motivate the children to learn their parents' skills and enhances their feeling of belonging.

At Sabyasachi, they also help preserve and promote cultural heritage by focusing on traditional Indian craftsmanship, benefiting the artisans, and contributing to the economic development of individuals involved in traditional crafts. Brands like Sabyasachi engage in models where artisans have some form of ownership or profit-sharing arrangements and empower artisans by giving them a stake in the brand's success.

The leadership of all four brands believes that the empowerment of the artisans and the farmers has strongly led to a change in the perception of the economic benefits of the art and skills of BOP individuals. Due to the brands' involvement, the BOP makes a consistent income, which takes away the seasonality of the occupation and is the first and most crucial denominator for the current generation of BOP to train future generations as well. Further, constant communication with other company stakeholders and positive consumer reviews provide a sense of achievement and belongingness to BOP, inspiring future generations to learn the skills.

To summarize, the brands must ensure that the unique skills of the farmers and artisans are passed on to future generations for the brand's long-term sustainability and BOP wellbeing. Thus, the brands also train the coming generations, and the current generations inspire them to learn the skillset for a better future.

8.5 Conclusion

The current chapter is about how luxury brands can create a positive social impact and work for the wellbeing of BOP individuals. The chapter suggests that the inclusion of BOP in the value chain of luxury brands as suppliers leads to better brand positioning and initiates BOP wellbeing. The chapter argues that accessibility to various resources alone cannot work for BOP wellbeing. Hence, the luxury brands need to do more for the mutual benefit. Therefore, the chapter suggests five different strategies that help attain different elements of BOP wellbeing along with the sustainable growth of the luxury brand. These strategies are accessibility to new markets, knowledge, and resources, blending of corporate and corporate elements of the company through collaboration, establishing end-to-end communication processes among the consumers, suppliers, and employees, educating consumers for mindful consumption through storytelling to increase the demand and tactics to preserve the unique skills of artisans and farmers for long term sustainability. These strategies resulted from analysing the processes of four Indian luxury brands that they have established over time for the inclusion of BOP. The tactics the brands apply are contextual and per the needs of the individuals at the base of the pyramid. The contextualization is done based on cultural and geographical differences. Additionally, the skill set that the artisans and farmers hone as per the industry to the brands belong.

References

Agyekumhene, C., De Vries, J. R., Van Paassen, A., Schut, M., & MacNaghten, P. (2020). Making smallholder value chain partnerships inclusive: Exploring digital farm monitoring through farmer friendly smartphone platforms. *Sustainability, 12*(11), 4580. https://doi.org/10.3390/su12114580

Alam, M. K. (2020). A systematic qualitative case study: Questions, data collection, NVivo analysis and saturation. *Qualitative Research in Organizations and Management an International Journal, 16*(1), 1–31. https://doi.org/10.1108/qrom-09-2019-1825

Atkinson, S. D., & Kang, J. (2021). New luxury: Defining and evaluating emerging luxury trends through the lenses of consumption and personal values. *Journal of Product & Brand Management, 31*(3), 377–393. https://doi.org/10.1108/jpbm-09-2020-3121

Bhattacharya, S., & Dalal, A. (2024). Exploring the brand gratitude paradigm at BOP in the context of emerging markets. *International Review on Public and Nonprofit Marketing.* https://doi.org/10.1007/s12208-024-00403-y

Birch, K. (2024). *Top 10: Sustainability strategies of luxury fashion brands.* Sustainabilitumag.com.

Chmielewski, D. A., Dembek, K., & Beckett, J. R. (2018). 'Business unusual': Building BoP 3.0. *Journal of Business Ethics, 161*(1), 211–229. https://doi.org/10.1007/s10551-018-3938-7

Cristini, H., Kauppinen-Räisänen, H., & Woodside, A. G. (2022). Broadening the concept of luxury: Transformations and contributions to well-being. *Journal of Macromarketing, 42*(4), 673–685. https://doi.org/10.1177/02761467221116779

Debenedetti, A., Philippe, D., & Dion, D. (2024). Balancing exclusivity and inclusivity through the strategic domestication of the luxury retail experience. *International Journal of Research in Marketing.* https://doi.org/10.1016/j.ijresmar.2024.05.001

Dembek, K., Sivasubramaniam, N., & Chmielewski, D. A. (2019). A systematic review of the bottom/base of the pyramid literature: Cumulative evidence and future directions. *Journal of Business Ethics, 165*(3), 365–382. https://doi.org/10.1007/s10551-019-04105-y

Derks, M., Oukes, T., & Romijn, H. (2022). Scaling inclusive business impacts at the base of the pyramid: A framework inspired by business model ecosystems research. *Journal of Cleaner Production, 366*, Article 132875. https://doi.org/10.1016/j.jclepro.2022.132875

Filho, W. L., Azeiteiro, U., Alves, F., Pace, P., Mifsud, M., Brandli, L., Caeiro, S. S., & Disterheft, A. (2017). Reinvigorating the sustainable development research agenda: The role of the sustainable development goals (SDG). *International Journal of Sustainable Development & World Ecology, 25*(2), 131–142. https://doi.org/10.1080/13504509.2017.1342103

Follman, J. (2012). BoP at ten: Evolution and a new lens. *South Asian Journal of Global Business Research, 1*(2), 293–310. https://doi.org/10.1108/20454451211252787

Ghosh, S., & Rajan, J. (2019). The business case for SDGs: An analysis of inclusive business models in emerging economies. *International Journal of Sustainable Development & World Ecology, 26*(4), 344–353. https://doi.org/10.1080/13504509.2019.1591539

Goal 8 | Department of Economic and Social Affairs. (n.d.). https://sdgs.un.org/goals/goal8

Gupta, S., & Kanungo, R. P. (2022). Financial inclusion through digitalisation: Economic viability for the bottom of the pyramid (BOP) segment. *Journal of Business Research, 148*, 262–276. https://doi.org/10.1016/j.jbusres.2022.04.070

Hanson, S., Jiang, L., Ye, J., & Murthy, N. (2018). Society or the environment? Understanding how consumers evaluate brand messages about corporate social responsibility activities. *Journal of Brand Management, 26*(1), 21–34. https://doi.org/10.1057/s41262-018-0110-8

Haworth, J., & Hart, G. (2007). Well-being—Individual, community and social perspectives. In *Palgrave Macmillan eBooks*. http://ci.nii.ac.jp/ncid/BB0883784X

Husain, R., Samad, T. A., & Qamar, Y. (2021). Past, present and future of luxury brands: A review and bibliometric analysis. *Journal of Fashion Marketing and Management, 26*(4), 582–602. https://doi.org/10.1108/jfmm-02-2021-0046

Ibrahim, N. F., Sharif, S. M., Saleh, H., Hasan, N. H. M., & Jayiddin, N. F. (2023). PERMA wellbeing and innovative work behaviour: A systematic literature review. *F1000Research, 12*, 1338. https://doi.org/10.12688/f1000research.141629.1

Jaén, M. H., Reficco, E., & Berger, G. (2020). Does integrity matter in BOP ventures? The role of responsible leadership in inclusive supply chains. *Journal of Business Ethics, 173*(3), 467–488. https://doi.org/10.1007/s10551-020-04518-0

Joy, A., Sherry, J. F., Venkatesh, A., Wang, J., & Chan, R. (2012). Fast fashion, sustainability, and the ethical appeal of luxury brands. *Fashion Theory, 16*(3), 273–295. https://doi.org/10.2752/175174112x13340749707123

Kang, E. Y., & Sung, Y. H. (2021). Luxury and sustainability: The role of message appeals and objectivity on luxury brands' green corporate social responsibility. *Journal of Marketing Communications, 28*(3), 291–312. https://doi.org/10.1080/13527266.2021.1874482

Karaosman, H., Perry, P., Brun, A., & Morales-Alonso, G. (2018). Behind the runway: Extending sustainability in luxury fashion supply chains. *Journal of Business Research, 117*, 652–663. https://doi.org/10.1016/j.jbusres.2018.09.017

Keller, K. L. (2017). Managing the growth tradeoff: Challenges and opportunities in luxury branding. In *Springer eBooks* (pp. 179–198). https://doi.org/10.1007/978-3-319-51127-6_9

Ko, E., Costello, J. P., & Taylor, C. R. (2017). What is a luxury brand? A new definition and review of the literature. *Journal of Business Research, 99*, 405–413. https://doi.org/10.1016/j.jbusres.2017.08.023

Kohnová, L., & Salajová, N. (2023). Impact of Industry 4.0 on companies: Value chain model analysis. *Administrative Sciences, 13*(2), 35. https://doi.org/10.3390/admsci13020035

Kolk, A., Rivera-Santos, M., & Rufín, C. (2013). Reviewing a decade of research on the "base/bottom of the pyramid" (BOP) concept. *Business & Society, 53*(3), 338–377. https://doi.org/10.1177/0007650312474928

Kumar, A., Kumra, R., & Singh, R. (2021). Base of the pyramid producers' constraints: An integrated review and research agenda. *Journal of Business Research, 140*, 115–129. https://doi.org/10.1016/j.jbusres.2021.11.046

Kumar, V., Khan, I., Fatma, M., & Singh, A. (2022). Engaging luxury brand consumers on social media. *Journal of Consumer Marketing, 39*(1), 121–132. https://doi.org/10.1108/jcm-10-2020-4175

Kunz, J., May, S., & Schmidt, H. J. (2020). Sustainable luxury: Current status and perspectives for future research. *BuR—Business Research, 13*(2), 541–601. https://doi.org/10.1007/s40685-020-00111-3

Lappeman, J., Ransome, K., & Louw, Z. (2019). Not one segment: Using global and local BoP characteristics to model country-specific consumer profiles. *European Business Review, 31*(3), 317–336. https://doi.org/10.1108/ebr-01-2018-0027

Lashitew, A. A., Bals, L., & Van Tulder, R. (2018). Inclusive business at the base of the pyramid: The role of embeddedness for enabling social innovations. *Journal of Business Ethics, 162*(2), 421–448. https://doi.org/10.1007/s10551-018-3995-y

Lee, Z., Alwi, S. F. S., & Gambetti, R. (2024). The thousand faces of beauty: How credible storytelling unlocks disability representation in inclusive luxury fashion branding. *Journal of Business Research, 181*, Article 114744. https://doi.org/10.1016/j.jbusres.2024.114744

London, T. (2008). The base-of-the-pyramid perspective: A new approach to poverty alleviation. *Academy of Management Proceedings, 2008*(1), 1–6. https://doi.org/10.5465/ambpp.2008.33716520

London, T., Anupindi, R., & Sheth, S. (2009). Creating mutual value: Lessons learned from ventures serving base of the pyramid producers. *Journal of Business Research, 63*(6), 582–594. https://doi.org/10.1016/j.jbusres.2009.04.025

Manchanda, P., Arora, N., Nazir, O., & Islam, J. U. (2023). Cultivating sustainability consciousness through mindfulness: An application of theory of mindful-consumption. *Journal of Retailing and Consumer Services, 75*, Article 103527. https://doi.org/10.1016/j.jretconser.2023.103527

Maxwell, J. (2005). *Qualitative research design.* Sage Publications.

Moore, J., Hanson, W., & Maxey, E. (2022). Beyond inclusion: Team impact of hiring people with disabilities at sephora and walgreens. *Organizational Development Journal, 40*(3).

Nadanyiova, M., & Das, S. (2020). Millennials as a target segment of socially responsible communication within the business strategy. *Littera Scripta, 13*(1). https://doi.org/10.36708/littera_scripta2020/1/8

Nielsen, C., & Aagaard, A. (2021). The fifth stage of business model research: The role of business models in times of uncertainty. *DOAJ (DOAJ: Directory of Open Access Journals).* https://doi.org/10.5278/jbm.v9i1.4246

Olšanová, K., Ríos, A. E., Cook, G., Král, P., & Zlatić, M. (2021). Impact of the awareness of brand-related CSR activities on purchase intention for luxury brands. *Social Responsibility Journal, 18*(3), 597–618. https://doi.org/10.1108/srj-10-2020-0398

Porter, M. (1985). *The competitive advantage: Creating and sustaining superior performance.* The Free Press.

Porter, M. E., & Kramer, M. R. (2007). Strategy and society: The link between competitive advantage and corporate social responsibility. *Strategic Direction, 23*(5). https://doi.org/10.1108/sd.2007.05623ead.006

Prahalad, C. K. (2011). Bottom of the pyramid as a source of breakthrough innovations. *Journal of Product Innovation Management, 29*(1), 6–12. https://doi.org/10.1111/j.1540-5885.2011.00874.x

Rahman, S. A., Amran, A., Ahmad, N. H., & Khadijeh Taghizadeh, S. (2013). Organizational supports to business success: Ensuring the wellbeing of BoP entrepreneurs. *International Journal of Conceptions on Management and Social Sciences, 1*(1).

Ruggiano, N., & Perry, T. E. (2019). Conducting secondary analysis of qualitative data: Should we, can we, and how? *Qualitative Social Work, 18*(1), 81–97. https://doi.org/10.1177/1473325017700701

Ranjan, K. R., Rohit, S., Dash, R., & Singh, R. (2021). Thinking, feeling and coping by BoP healthcare consumers: Policy-based intervention in an emerging market. *Journal of Marketing Management, 37*(9–10), 914–961. https://doi.org/10.1080/0267257x.2020.1862279

Reficco, E., & Márquez, P. (2009). Inclusive networks for building BOP markets. *Business & Society, 51*(3), 512–556. https://doi.org/10.1177/0007650309332353

Reyes-Menendez, A., Palos-Sanchez, P., Saura, J. R., & Santos, C. R. (2021). Revisiting the impact of perceived social value on consumer behavior toward luxury brands. *European Management Journal, 40*(2), 224–233. https://doi.org/10.1016/j.emj.2021.06.006

Rishi, M., Jauhari, V., & Joshi, G. (2015). Marketing sustainability in the luxury lodging industry: A thematic analysis of preferences amongst the Indian transition generation. *Journal of Consumer Marketing, 32*(5), 376–391. https://doi.org/10.1108/JCM-09-2014-1155

Roberson, Q., & Perry, J. L. (2021). Inclusive leadership in thought and action: A thematic analysis. *Group & Organization Management, 47*(4), 755–778. https://doi.org/10.1177/10596011211013161

Rosenbaum, M. S., Ramirez, G. C., Campbell, J., & Klaus, P. (2019). The product is me: Hyper-personalized consumer goods as unconventional luxury. *Journal of Business Research, 129*, 446–454. https://doi.org/10.1016/j.jbusres.2019.05.017

Rosenstock, T. S., Lubberink, R., Gondwe, S., Manyise, T., & Dentoni, D. (2020). Inclusive and adaptive business models for climate-smart value creation. *Current Opinion in Environmental Sustainability, 42*, 76–81. https://doi.org/10.1016/j.cosust.2019.12.005

Schoneveld, G. C. (2020). Sustainable business models for inclusive growth: Towards a conceptual foundation of inclusive business. *Journal of Cleaner Production, 277*. Article 124062. https://doi.org/10.1016/j.jclepro.2020.124062

Shukla, Y., Singh, R., Dwivedi, P., & Chatterjee, R. (2023). Wellbeing implications of BoP marketing: A service ecosystem approach. *Journal of Services Marketing, 37*(7), 883–894. https://doi.org/10.1108/jsm-03-2022-0105

Sinha, P. K., Gupta, S., & Rawal, S. (2017). Brand adoption by BoP retailers. *Qualitative Market Research an International Journal, 20*(2), 181–207. https://doi.org/10.1108/qmr-07-2014-0056

Staff, K. A. W. (2021, March 23). *How can the luxury industry become more inclusive?* Knowledge at Wharton. https://knowledge.wharton.upenn.edu/article/how-can-the-luxury-industry-become-more-inclusive/

Steptoe, A., Deaton, A., & Stone, A. A. (2014). Subjective wellbeing, health, and ageing. *The Lancet, 385*(9968), 640–648. https://doi.org/10.1016/s0140-6736(13)61489-0

Viswanathan, M., Sridharan, S., & Ritchie, R. (2009). Understanding consumption and entrepreneurship in subsistence marketplaces. *Journal of Business Research, 63*(6), 570–581. https://doi.org/10.1016/j.jbusres.2009.02.023

Viswanathan, M., Sreekumar, A., & Gau, R. (2018). Evolving and expanding marketing to address challenges and opportunities in BoP markets: Looking back and forward. In R. Singh (Ed.) *Bottom of the pyramid marketing: Making, shaping and developing BoP markets* (*Marketing in Emerging Markets*) (pp. 3–27). Emerald Publishing Limited. https://doi.org/10.1108/978-1-78714-555-920181002

Amrita Bansal is Assistant Professor of Marketing at the International Management Institute New Delhi, India. She has a total work experience of more than six years in reputable business schools of India. Her scholarly contributions manifest in publications in esteemed journals such as the Journal of Business Research, Journal of Promotion Management, Journal of Marketing Communication. Further, she has also presented her work at international conferences like IABS conference by 'Business and Society Journal at University of Bath, UK and International Conference on Corporate and Marketing Communications. Her research interests drive into themes ranging from gender portrayal in advertisements to co-creating content through product storytelling to sustainable brands.

Varsha Jain is the AGK chair professor of marketing and co-chairperson of research and dissertation at the MICA, India. She is the deputy editor of the Journal of Consumer Behaviour (JCB, "A" Journal) and an associate editor (AE) at three "A" journals: Psychology and Marketing, the Journal of Consumer Marketing, and the Australasian Marketing Journal. She has authored over 180+ publications (primarily qualitative research in the digital age), including multiple papers in A* and A journals, including the European Journal of Marketing, the International Journal of Information Management, the Journal of Advertising Research, the Journal of Consumer Behavior, the Journal of Consumer Marketing, and the Journal of Business Research.

Chapter 9
Sustainable Consumer Behaviour in Emerging Markets

Mahima Mathur

Abstract The term sustainable consumer (SC) behaviour is recognised as the extent to which consumption decisions are made with the goal of minimising or improving the effects on society and the environment. The global market is often thought to be socially and environmentally aware of how humans affect the environment. However, responses and expectations of consumers may vary depending on contextual variables of their region. For example, while individuals at the BoP markets in emerging economies are concerned about the environment, questions regarding prices and personal benefits may override when making purchase decisions. In this chapter we discuss the dimensions and complexities of sustainable consumer (SC) behaviour including segmentation characteristics, SC predictors and willingness to pay with a special focus on consumers in emerging markets that are largely operating at the BoP. Thereafter, we propose a comprehensive description of a SC by bifurcating pertinent and relevant factors that are known to be helpful in predicting SC behaviour among internal or external influences on the individual (actor) or behaviour (actions). At the end, we elucidate few recommendations for practitioners and policymakers. The description of a SC will be useful to environment and social marketers who wish to target consumers and design intervention strategies for this target group, and researchers who wish to expand SC studies.

Keywords Bottom of the pyramid · Base of the pyramid · Sustainable consumption · Socially responsible consumer behaviour · Sustainability · ESG

M. Mathur (✉)
Department of Retail Management, Prin. L. N. Welingkar Institute of Management Development & Research, Mumbai, India
e-mail: mahima.mathur@welingkar.org

R. Singh and T. A. Wani (eds.), *Marketing to the Base of the Pyramid*,
https://doi.org/10.1007/978-981-96-7105-2_9

9.1 Introduction

The sustainability movement that began in the developed world decades ago is now intensifying in the emerging markets. Companies operating in these markets are facing increasing pressure from consumers, trading partners, investors, regulators, and employees to support their commitments with demonstrated action. Recent reports suggest that the sustainability imperative is rapidly changing the nature of the global business landscape as companies realise that scores in environmental, social and governance (ESG) metrics are not only important for their performance, but also to mitigate overall risk and disruption (Tansan et al., 2023). The bottom of the pyramid consumers (BoP), comprising the largest (approximately 4 to 5 billion) and the poorest socio-economic segment, reside mainly in emerging markets (Hammond & Prahalad, 2004; London & Hart, 2004). These consumers at the BoP, sustaining on an annual per capita income of less than $3,000, live in 164 developing and 47 least developing countries such as India, Srilanka, Brazil, China, Bangladesh and Zimbabwe (Srivastava et al., 2020).

Consumer behaviour of the BoP segment is unique because of their inhibited circumstances. BoP customers are more vulnerable than mainstream consumers because of their extended exposure to poverty because of illiteracy, inequality, social and economic constraints, and executive function impairments (Chakravarti, 2006; Mathur et al., 2020; Mishra et al., 2021). Literature in recent past examines concepts such as image consciousness, aspirational consumption, compensatory consumption, social comparison and collectivist culture with respect to purchase decisions of the heterogeneous set of consumers at the BoP (Jaiswal & Gupta, 2015; Srivastava et al., 2020).

It is commonly believed that the global market is socially and environmentally conscious of the impact that humans have on the environment. Hence, it is believed that consumers will demand social and environmental responsibility from companies, irrespective of their country of origin and location (Abugre & Nyuur, 2015). However, responses and expectations of consumers may vary depending on contextual variables of their region (Rathod et al., 2022). The term sustainable consumer (SC) behaviour is recognised as the extent to which consumption decisions are made with the goal of minimising or improving the effects on society and the environment (Trudel, 2019). SCs are concerned about the environment and social welfare while satisfying personal needs.

The broad approaches used to segment these set of consumers are socio-demographic characteristics, environmental values and psychological factors (Akehurst et al., 2012; Anderson & Cunningham, 1972). However, socio-demographic characteristics such as sex, age, region, nation, especially income and education have been found to be inconsistent while predicting SC behaviour (Ali et al., 2021; Elhoushy & Lanzini 2021). It was found that factors such as perceived ability to reduce threat due to knowledge and awareness of consequences (Bouman et al., 2020), values such as collectivism (Kim & Choi, 2005) and pro-environmental attitude (Naatu et al., 2023) are more important in explaining SC behaviour. It is also important to note that SC behaviour only indicates a predisposition that consumers care about the environment, but their buying behaviour or willingness to buy may not be the same (Barbaro-Forleo, 2001; Gardyn, 2003).

In this chapter we discuss the dimensions and complexities of sustainable consumer (SC) behaviour including segmentation characteristics, SC predictors and willingness to pay with a special focus on consumers in emerging markets that are largely operating at the BoP. The discussion aims to use evidence from literature to examine the SC behaviour of the BoP consumers in emerging markets to assist companies to meet sustainability goals while operating in these markets. Towards the end of the chapter, we propose a comprehensive description of a SC by bifurcating pertinent and relevant factors that are known to be helpful in predicting SC behaviour among internal or external influences on the individual (actor) or behaviour (actions).

9.2 Consumer Behaviour of the BOP Segment

The BoP market is characterised with irregular and low incomes, geographically dispersed population that is relatively isolated with scarce resources, poor health conditions, low education levels and are mainly occupied in the informal sector while residing in rural or slum areas (Mathur et al., 2020; Prahalad, 2004; Venugopal & Vishwanathan, 2017). The BoP consumers are not a homogeneous segment with the same buying power, instead they are motivated to purchase products for varied reasons and have diverse socio-demographic characteristics that effect their purchase decisions (Jagadale et al., 2018; Sheth, 2011). They are also considered to be more price sensitive than mainstream consumers and prefer products that are explicitly tailored to meet their needs (Basu et al., 2013). It is for these reasons that the consumer behaviour patterns at the BoP are unique, and the success of companies lies in adapting their business models around them.

The main patterns of purchase behaviour of the consumers at the BoP that have been examined by studies in the past are summarised in Table 9.1.

The theory of social comparison and compensatory consumption was found to be most prevalent in describing purchase behaviour of BoP markets such as India, Brazil, Bangladesh and Zimbabwe (Kolk et al., 2014). It was also found that some of these theories may lead to certain positive outcomes for the BoP consumer segment. For example, social comparison may result in increased spending on children's education as a response to social-exclusion and socio-economic marginalisation by the reference groups, aspirational consumption may lead to aspiring for decent shelter and better education for children, it may also result in purchasing sustainable products mimicking the top-of-the-pyramid market. These theories explain the peculiar consumer behaviour at the BoP.

Consumers across all segments are increasingly becoming conscious of the impact of their consumption-related choices on the environment and society at large (Kleanthous, 2011). The next section in the chapter focuses on studying relevant

Table 9.1 Patterns of purchase behaviours at the BoP

Concept/Theory	Description	Sample studies
Compensatory consumption	Using purchasing behaviour as a substitute to fulfil the lack of satisfaction of needs such as social-power, social-status, social-inclusion, self-esteem, etc	Jaiswal and Gupta (2015), Mukherjee et al. (2020), Zheng et al. (2018)
Social comparison	Customers may buy goods that will promote social inclusion if they experience discrimination or social exclusion, such as second-hand luxury products It may trigger aspirational-consumption or compensatory-consumption	Mukherjee et al. (2020), Srivastava et al. (2020)
Aspirational consumption	An individual will spend money on status-symbol products and services to match the reference group's consumption It also gives rise to improved well-being	Duffy (2016), Jaikumar et al. (2018)
Second-hand purchase	Acquiring used goods for reasons such as affordability, durability, accessibility, sustainability, and social status	Arunachalam et al. (2019), Mukherjee et al. (2020)
Image-consciousness	An external disposition in individuals of being concerned about their appearance and impression on others while making purchase decisions. It aligns to the extrinsic goal of fame	Gupta and Srivastav (2016), Srivastava et al. (2020)
Collectivist culture with respect to consumption	A cultural pattern where individuals prioritise group harmony, interdependence and conformity to societal expectations and norms while making consumption-related choices	Chikweche and Fletcher (2014), Jaiswal and Gupta (2015)

Source Authors

aspects of sustainable consumer (SC) behaviour, segmentation profile, individual and group-level predictors and willingness to pay more for sustainable products.

9.3 Sustainable Consumer Behaviour

The terms SC and SC behaviour are often used synonymously with terms such as "socially responsible consumer" (Berkowitz & Lutterman, 1968), "green consumer" (Gilg et al., 2005), "ecologically conscious consumer" (Roberts & Bacon, 1997), "pro-environmental behaviour" (Abdul-Muhmin, 2007) and "responsible sustainable consumer behaviour" (Hosta & Zabkar, 2021). SC is someone whose consumption decisions are made with the goal of minimising or improving the effects on society and the environment (Anderson & Cunningham, 1972; Trudel, 2019). Hence, they are concerned about the environment and social welfare while satisfying personal needs. The research studies, while examining SC behaviour, argue that because consumers are an integral part of the ecosystem they are affected not only by their own cognitive influences and personal characteristics, but also by interactions with their social and physical environments (Lockrey et al., 2018; Moschis et al., 2020; Phan-Le et al., 2024). Hence, SCs cannot act without the influences in the individual's ecosystem and that systemic factors will affect both consumption and behaviour.

Few theories around categorising SCs report for behaviours (such as purchasing sustainable products), assessing consumers' personal characteristics (such as age, income, gender, education etc.), or involve more subjective aspects such as attitudes and worldviews towards the environment. The three broad approaches used to categorise the SC segment are: (1) socio-demographic characteristics, (2) environmental concerns and values, and (3) psychological-values (Anderson & Cunningham, 1972; Berkowitz & Lutterman, 1968). A combination of these approaches is often used to define SC's behaviour. For example, customers who want to meet their own needs but are also mindful of the welfare of the environment and society can be defined as "socio-psychological". Although, the underlying assumption is that attitudes will lead to intention-to-purchase which will lead to behaviour. However, the correlation between intention and behaviour is still unclear for sustainability (Rausch & Kopplin, 2021; Zaremohzzabieh et al., 2021). The three approaches of segmenting SCs are explained below in detail:

Socio-demographic characteristics—Socio-demographic variables such as age, income, gender, region, nation, etc., are relatively easy to observe, as they are objective. Since these characteristics are fixed and cannot be influenced by any interventions, marketers can only adopt marketing strategies accordingly (Brennan et al., 2016). However, socio-demographic characteristics have not been found to be consistent in predicting SC's behaviour (Ali et al., 2021; Elhoushy & Lanzini, 2021). For example, there are empirical studies which confirm that women are more likely to exhibit environmental consciousness in their purchase behaviour (Kassinis et al., 2016), while others show that men exhibit stronger relationship than women (Chun et al., 2021). Income has usually been seen as positively correlated with SC behaviour, studies conducted on this variable do not support this view (Anderson & Cunningham, 1972). Similarly, while many studies found that consumers with higher education are more likely to exhibit SC behaviour (Roberts, 1996; Zimmer et al., 1994). Straughan and Roberts (1999) examined that literacy did not have a positive correlation with SC behaviour.

Environmental concerns and values—Environmental concerns and values reflect an individual's knowledge of environmental issues and readiness to contribute to their solution (Dunlap & Jones, 2002). It is a method of psychographic segmentation that investigates a person's behaviours, interests, opinions, beliefs, and values to determine the reasons behind their actions. This approach examines the "why" and "what" aspect of segmentation whereas the demographic approach focuses on the "who" and "where". The environmental concerns construct discusses two social-value dimensions: "altruistic-egoistic" and "conservative-open to change" (Weigel & Weigel, 1978). The "altruistic-egoistic" dimension refers to pro-social versus pro-self behaviour of an individual. While the "conservative-open to change" dimension refers to the closed-mindedness versus open-mindedness in individuals. Researchers have found a positive relationship between environmental concerns and SC behaviour (Gifford & Nilsson, 2014; Hoffmann et al., 2022).

Another related well-researched area in SC studies is "worldviews towards the environment" that provides direction towards SC behaviour (Dunlap, 2008). The individual's worldviews are often not directly observable but indicate the way they approach life and issues in the world. They are formed because of their cultural and personal experiences (Brennan et al., 2014). It was found that active environmental activism has been linked to individuals who value the natural environment greatly in their lives (Gong & Chu, 2022). The New Ecological Paradigm (NEP) scale is one of the initial models to measure worldview with respect to the environment. It includes measurement of an individual's values, beliefs, norms and environmental concerns (Brennan et al., 2014).

Psychological values—Personal attitudes and beliefs regarding their own sustainable consumption behaviour held by individuals is a commonly used approach to characterise SCs. This approach relies on the psychographic variables such as altruism, perceived-consumer-effectiveness (PCE), and self-efficacy (Akehurst et al., 2012). Notably, most definitions of SC adhere to the internal environmental focus of control (INELOC) model. According to the INELOC model, the SC makes decisions in a free-choice, individualised manner. It suggests that the SC makes a deliberate decision between products that are environmentally friendly and those that are not (Brennan et al., 2016). However, extrinsic elements are not considered as contributing factors to an individual's behaviour in INELOC model-based studies (Hwang et al., 2021).

Altruism—Altruism has been defined as having a community-oriented outlook by consumers when faced with purchase-related decisions (Antil, 1984). It plays a crucial role in political activism and has a significantly positive relationship with SC behaviour (Straughan & Roberts, 1999).

Perceived consumer effectiveness (PCE)—PCE is defined as the individual's assessment of the degree to which their actions such as recycling, purchasing sustainable products, e-invoices, etc., can make a difference in resolving environmental and social concerns (Ellen et al., 1991). It is found that individuals tend to behave more proactively if they believe their actions will be effective in solving environmental problems (Moisander, 2007). PCE is also shown to be more relevant than other socio-demographic and psychographic variables, and important in predicting SC behaviour (Roberts & Bacon, 1997; Straughan & Roberts, 1999).

Self-efficacy—Self-efficacy (SE) is related to the individual's belief in their ability to perform the task of making a difference in solving environmental and social concerns. It has been found to impact the individual's goals, effort and likelihood of success in sustainable behaviour (Lee et al., 2019).

Although social, economic, and demographic traits have been used to identify the SC characteristics, several researchers contend that psychographic factors such as environmental concerns and values, and psychological values offer more pertinent insights into the behaviour of SCs (Anderson & Cunningham, 1972; Awad, 2011; Chan, 1999).

Collectivism—An additional value that influences SC behaviour is collectivism (Triandis, 1993). Collectivism is a type of human culture that implies consideration, cooperation, and prioritisation of a group over an individual. A collectivist may sacrifice personal motivations for the benefit of the group as opposed to an individualist who is more focused on depending only upon himself or herself and often competes with others for status (McCarty & Shrum, 1994). On the other hand, individualism is a measure of how much a person depends on their own independence. Even when they are part of a community or group, individualist consumers maintain their individuality. They also depend more on their own accomplishments than on belonging to a group, and they are known to compete with others for social prestige (Triandis, 1993). Individualists are typically more hostile to the environment, while collective thinkers are known to be more environmentally friendly (Triandis, 1993). Research also suggests that the value of fun and enjoyment was positively correlated to SC-related behaviour due to the fulfillment an individual may feel through interaction with the environment (McCarty & Shrum, 1994).

Attitudes—Attitude is a psychological tendency that is demonstrated by evaluating something favorably or unfavourably (Brennan et al., 2014). They have considerable influence in motivating sustainable behaviours. The most researched attitudes in the SC behaviour literature are of perceived importance and inconvenience. The attitude of perceived importance is the degree to which individuals show concern about environmental issues and whether they view their environmentally compatible behaviour of significance to themselves and society (Amyx et al., 1994). The attitude of inconvenience focuses on how inconvenient an individual thinks it is to act in a way that is environmentally friendly. For example, an individual may feel that single-use plastic is bad for the environment, but still use them as they are convenient. Similarly, an individual may know the benefits of recycling for society but may find it personally inconvenient to adopt. McCarty and Shrum (1994) found that regardless of how important they thought recycling was, individuals were less inclined to adopt recycling if they felt it was inconvenient. Hence, it appears that perceived inconvenience has a greater influence on the individual's actions regardless of the importance of their beliefs.

Another aspect of attitude that is important in the context of SC behaviour is the attitude-to-behaviour consistency which means the alignment of attitude with behaviour (Ajzen, 2011). Most researchers in the attitude-to-behaviour consistency domain regard behaviour from the "rational" choice viewpoint, from the SC perspective it means choosing environmentally friendly products that balance the individual's and environment's needs (Ziesemer et al., 2021). However, results from Litvine and Wüstenhagen (2011) show that attitudes are subordinate to other factors affecting SC behaviour such as perceived-consumer-effectiveness, self-efficacy, psychological values and other personal norms. This explains the situation where an individual may have a favourable attitude towards the environment but continue to behave unsustainably.

Behaviours—It has been found that many behaviours are driven by routines and habits that are learned through repetition instead of a mindful consideration of values and beliefs (Verplanken et al., 2018). Hence, identifying the behavioural patterns of individuals, both habitual and considered, will be important in describing a SC. As per definition, an SC will be prepared to change their existing habits and routine in favour of making sustainable decisions. This will involve the ability to reflect on their existing habits, then suspend the habit to adopt a new routine that is more sustainable (Wood & Rünger, 2016). There are two theories involved in the disruption of habits; the habit-discontinuity hypotheses and the self-activation hypotheses (Verplanken & Orbell, 2003; Verplanken et al., 2008). The habit-discontinuity hypotheses posit that when a change in context results in disruption of old habits, consumers are more likely to consciously consider changing their behaviour (Verplanken et al., 2018). The self-activation hypothesis posits that values can guide behaviour when they are either embedded in the self-concept or are consciously activated by changes in the context (Verplanken et al., 2008). These changes in context could be on account of important life events or developmental evolution. Thus, behavioural interventions can be constructed to disrupt habits and enable the development of new routines, resulting in reduced dependence on the individual's cognitive and affective engagement towards SC behaviour (Lockrey et al., 2023).

9.4 Comprehensive Description of SC

After a thorough analysis of the relevant literature, various factors that describe a SC and are helpful in predicting SC behaviour have been characterised. These factors can be categorised into (1) psychographic, (2) psychological, (3) socio-cultural and (4) behavioural. While most current definitions of SC are one-dimensional and research an individual's (actor) characteristic or their behaviour (actions) separately. A SC's environment is influenced by a number of internal and external elements that lead to sustainable consumption (Phan-Le et al., 2024). Even though there is no formal method for selecting pertinent segmentation variables, it is important to understand the context of behaviour of SCs before the variables can be selected to identify meaningful segments (Kubacki et al., 2017). A comprehensive description of the SC, where the four factors namely, psychographic, psychological, socio-cultural and

	Actor	Action
Internal	Psychological - Altruism - Perceived-consumer-effectiveness - Self-efficacy - Attitudes o Perceived importance o Inconvenience	Behavioural - Routines and habits o Habit discontinuity o Self-activation
External	Psychographic - Environmental concerns and values - Worldviews towards environment	Socio-cultural - Collectivism - Fun and enjoyment

Fig. 9.1 Comprehensive description of a SC (*Source* Authors)

behavioural have been distinguished as internal or external factors and related to the actor or their actions, has been depicted in a matrix as shown in Fig. 9.1.

The proposed description of a SC enables a detailed understanding of the internal and external factors that impact SC behaviour. The psychological factors such as altruism, PCE, SE, and attitudes of perceived importance and inconvenience are internal to an individual, directly motivating SC's behaviour. However, they are unstable on their own as an individual's behaviour evolves over time because of external influences. The actions or behaviour of an individual, habitual and consciously considered, are determined by their self-concept and ability to activate change in a given context. Thus, the behaviour of change in routines and habits is an additional internal factor describing a SC. As discussed during the literature reviewed in the previous section, there are two external factors at the individual and behavioural level that play an important role in determining sustainable behaviour; firstly, psychographic characteristics of an individual such as environmental concerns and values and worldviews towards the environment, secondly, socio-cultural factors such as, collectivism and the element of fun and enjoyment. Environmental concerns are related to knowledge and emotions regarding environmental problems that are helpful in shaping the motives of SC behaviour. On the other hand, it is possible to map societal attitudes towards sustainability using worldviews. The socio-cultural factor collectivism is a cultural pattern where people prioritise group goals more than individual goals. It is seen that in such communities' pro-social concerns and altruistic generosity are highly valued motives guiding sustainable behaviour (Jiao & Zhao, 2023).

9.5 Sustainable Consumer Behaviour at the BOP

Although, some research supports the popular belief that culture, societies and the level of economic growth determines the individual's attitude and behaviour (Chu et al., 2020). For example, empirical studies in developed countries such as the USA and Europe suggest that consumers tend to exhibit higher social responsibility (Chu et al., 2020; Papadopoulou et al., 2022). However, while consumers in Brazil, China and India were found most concerned about climate change, consumers in the UK, USA and Germany are less concerned (Carrighan & Pelsmacker, 2009). The discussion in the previous section also indicates that the relationship between age, gender, income, education and SC behaviour was found to be unreliable (D'Souza et al., 2007; McEvoy, 1972; Zimmer et al., 1994). For instance, income was always thought to be positively correlated to SC behaviour, because of the popular belief that most sustainable products are more expensive than other products (Awad, 2011). However, the results of much research on this variable do not support this view (Anderson & Cunningham, 1972; Roberts & Bacon, 1997; Zimmer et al., 1994). Similarly, while many empirical studies found that consumers with higher levels of education are more susceptible to SC behaviour (Roberts, 1996; Zimmer et al., 1994). Straughan and Roberts (1999) stated that literacy did not have a positive relationship with SC behaviour. Hence, it can be concluded that psychographic factors such as environmental concerns, psychological values and collectivism culture are better indicators, while socio-demographic characteristics are less reliable in predicting overall SC behaviour.

Few studies have examined the response and expectations of individuals at the BoP market regarding sustainability (Abdul-Muhmin, 2007; Korsunova et al., 2022; Naatu et al., 2023). Their findings indicate that contrary to popular beliefs, the SC behaviour at the BoP market is not dependent on personal values that may be congruent to globally held beliefs but are influenced by their knowledge of environmental concerns because of their personal convictions and experiences regarding their ability to make a difference. It seems that the BoP market has not yet fully embraced the issues of global sustainability due to reasons such as network and infrastructure issues along with behavioural control or self-efficacy related problems (Zaki et al., 2014). In terms of SC behaviour, their perceptions to make a difference are more associated with environmental advocacy than purchase behaviour towards sustainable products. This explains the reason why people in BoP markets are more likely to engage in protests and demonstrations against organisations who they believe are a threat to their local environment. For example, young people in Ghana threatened to form a militia against oil producing companies when the locals realised a decline in fish harvest after the oil companies set up their operations in the region (Mohammed et al., 2022). While individuals at the BoP market may be concerned about the environment, questions regarding prices and personal benefits override when making purchase decisions.

In the low-income context of emerging economies, the limitation of resources often compels individuals to adopt innovative practices that conserve resources while retaining the value of material and goods in circulation (Korsunova et al., 2022). This is especially true for the informal sector, which makes up close to 50 percent of the economic output in some emerging markets (Lee et al., 2021). Thus, a section of literature in the area argues that unlike policy-driven adoption of SC behaviour in developed markets, in emerging markets SC behaviour is driven by economic necessity that is often reflected in the everyday practices of individuals at the BoP segment (Preston & Lehne, 2017). For example, reuse and repurpose strategies are adopted in India, Tanzania and Brazil by making packaging material out of scraps of different materials such as jute and leather.

9.6 Recommendations to Practitioners and Policymakers

The comprehensive description framework suggested in the chapter will guide practitioners in organisations trying to sell sustainable products, and social marketers and policymakers in designing campaigns to appeal to different levels of SCs. This means not just targeting based on demographic or behavioural factors but considering the psychological, psychographic and socio-cultural factors as well. For example, campaigns can target consumers who already exhibit a high degree of environmental concern and values. This kind of strategic targeting will ensure that the sustainability campaigns are more relevant and effective to the consumers.

Regarding the sustainable consumer behaviour at the BOP, the managers and policymakers should introduce a clear dimension of action to them while building strategies and campaigns, because there is evidence that it is important from a customer perspective to understand the impact of their actions and their ability to make a difference.

The chapter also highlighted that for consumers in low-income markets, the prices and personal benefits take precedence over the sustainable aspect of the product. Although, addressing price sensitivity while promoting sustainable products is a real challenge for companies. However, a few strategies such as highlighting the long-term value of the product in terms of its durability, longevity and cost savings from efficiency can make a difference. For example, Ikea's sustainable range of home furnishing products is promoted as stylish and durable. Another strategy that can be adopted by companies is to offer affordable entry-level products without compromising on the environmental benefits.

9.7 Conclusion

In this chapter we begin by examining the main patterns of purchase behaviour of the BoP consumers. The theories of social comparison and compensatory consumption exhibit the peculiar nature of consumer behaviour at the BoP. In the next section we start by defining the SC and establish that while social, economic and demographic traits such as age, income, sex, and education have been used to identify SC characteristics, psychographic factors such as environmental concerns and values, worldview towards the environment, altruism, perceived-consumer-effectiveness, self-efficacy, collectivism, etc., offer more better insights into the behaviour of SCs. Additionally, although an individual's personal factors have been the focus of a lot of research, it has been found to not be enough to create behavioural change in respect to the environment. The SC needs to be considered with complex internal and external systems which is difficult to define and describe with personal factors alone. Based on the findings after an exhaustive search of the literature we proposed a comprehensive description of a SC that provides a holistic understanding of sustainable behaviour. This description can be treated as a framework that will be useful for (i) environment and social-marketers who wish to target consumers showing an inclination for sustainable behaviour, (ii) social-marketers who are planning to design intervention strategies for this target group, and (iii) researchers who wish to further develop SC studies by understanding the inter-relationships between various internal and external factors describing a SC.

In the last section of this chapter, we discuss that while consumers across all segments are increasingly becoming conscious of the impact of their consumption-related choices on the environment and society at large, it seems that the BoP market is yet to fully connect with global sustainability challenges and their perceptions to make a difference are more associated to environmental advocacy than purchase behaviour towards sustainable products. When making purchase decisions the prices and personal benefits over-ride other aspects of the product. Lastly, SC behaviour in BoP markets, that are mainly found in emerging economies, is driven by economic necessity that is often reflected in the everyday practices of individuals.

References

Abdul-Muhmin, A. G. (2007). Explaining consumers' willingness to be environmentally friendly. *International Journal of Consumer Studies, 31*(3), 237–247.

Abugre, J. B., & Nyuur, R. B. (2015). Organizations' commitment to and communication of CSR activities: Insights from Ghana. *Social Responsibility Journal, 11*(1), 161–178.

Ajzen, I. (2011). The theory of planned behaviour: Reactions and reflections. *Psychology & Health, 2011*(26), 1113–1127.

Akehurst, G., Afonso, C., & Martins Gonçalves, H. (2012). Re-examining green purchase behaviour and the green consumer profile: New evidences. *Management Decision, 2012*(50), 972–988.

Ali, S., Akter, S., & Fogarassy, C. (2021). The role of the key components of renewable energy (Combustible renewables and waste) in the context of CO_2 emissions and economic growth of selected countries in Europe. *Energies, 14*, 2034.

Amyx, D. A., Dejong, P. F., Lin, X., Chakraborty, G., & Wiener, J. L. (1994). Influencers of purchase intentions for ecologically safe products: An exploratory study. In C. W. Park et al. (Eds.), *AMA winter educators' conference proceedings 5* (pp. 341–347). American Marketing Association.

Anderson, W. T., & Cunningham, W. (1972). The socially conscious consumer. *Journal of Marketing, 36*, 23–31.

Antil, J. H. (1984). Socially responsible consumers: Profile and implications for public policy. *Journal of Macromarketing, 4*(2), 18–39.

Arunachalam S., Bahadir, S. C., Bharadwaj, S. G., & Guesalaga, R. (2019). New product introductions for low-income consumers in emerging markets. *Journal of the Academy of Marketing Science*, 1–27

Awad, T. A. (2011). Environmental segmentation alternatives: Buyers' profiles and implications. *Journal of Islamic Marketing, 2*(1), 55–73.

Barbaro-Forleo. (2001). Targeting consumers who are willing to pay more for environmentally friendly products. *Journal of Consumer Marketing, 18*(6), 503–520.

Basu, R. R., Banerjee, P. M., & Sweeny, E. G. (2013). Frugal innovation: Core competencies to address global sustainability. *Journal of Management for Global Sustainability, 1*(2), 63–82.

Berkowitz, L., & Lutterman, K. G. (1968). The traditional socially responsible personality. *Public Opinion Quarterly, 32*, 169–185.

Bouman, T., Steg, L., & Zawadzki, S. J. (2020). The value of what others value: When perceived biospheric group values influence individuals' pro-environmental engagement. *Journal of Environmental Psychology, 71*, Article 101470.

Brennan, L., Binney, W., Aleti, T., & Parker, L. (2014). Why validation is important: An example using the NEP scales. *Mark. Soc. Res., 22*, 15–31.

Brennan, L., Previte, J., & Fry, M.-L. (2016). Social marketing's consumer myopia. *Journal of Social Marketing, 6*, 219–239.

Carrighan, M., & Pelsmacker, P. (2009). Will ethical consumers sustain their values in the global credit crunch? *International Marketing Review, 26*(6), 674–687.

Chakravarti, D. (2006). Voices unheard: The psychology of consumption in poverty and development. *Journal of Consumer Psychology, 16*(4), 363–376.

Chan, K. (1999). Market segmentation of green consumers in Hong Kong. *Journal of International Consumer Marketing, 12*(2), 7–24.

Chikweche, T., & Fletcher, R. (2014). Marketing to the "Middle of the Pyramid". In K. Perks & D. P. Tej Adidam, (Eds.), *Emerging Markets Using a Social Network Perspective: Evidence from Africa*. (Vol. 9, pp. 400–23). Emerald Group Publishing. https://doi.org/10.1108/IJoEM-05-2013-0090

Chu, S. C., Chen, H. T., & Gan, C. (2020). Consumers' engagement with corporate social responsibility (CSR) communication in social media: Evidence from China and the United States. *Journal of Business Research, 110*, 260–271.

Chun, E., Joung, H., Lim, Y. J., & Ko, E. (2021). Business transparency and willingness to act environmentally conscious behavior: Applying the sustainable fashion evaluation system "Higg Index." *Journal of Global Scholars of Marketing Science, 31*, 437–452.

D'Souza, C., Taghian, M., & Khosla, R. (2007). Examination of environmental beliefs and its impact on the influence of price, quality and demographic characteristics with respect to green purchase intention. *Journal of Targeting, Measurement and Analysis for Marketing, 15*(2), 69–78.

Duffy, B. E. (2016). The romance of work: Gender and aspirational labour in the digital culture industries. *International Journal of Cultural Studies, 19*(4), 441–457.

Dunlap, R. E. (2008). The new environmental paradigm scale: From marginality to worldwide use. *The Journal of Environmental Education, 2008*(40), 3–18.

Dunlap, R. E., & Jones, R. E. (2002). Environmental concern: Conceptual and measurement issues. In *Handbook of environmental sociology* (vol. 3, pp. 482–524).

Ellen, P. S., Wiener, J. L., & Cobb-Walgren, C. (1991). The role of perceived consumer effectiveness in motivating environmentally conscious behaviors. *Journal of Public Policy & Marketing, 10*, 102–117.

Elhoushy, S., & Lanzini, P. (2021). Factors affecting sustainable consumer behavior in the MENA Region: A systematic review. *Journal of International Consumer Marketing, 33*, 256–279.

Gardyn, R. (2003). Eco-friend or foe? *American Demographics, 25*(8), 12.

Gifford, R., & Nilsson, A. (2014). Personal and social factors that influence pro-environmental concern and behaviour: A review. *International Journal of Psychology, 49*, 141–157.

Gilg, A., Barr, S., & Ford, N. (2005). "Green consumption or sustainable lifestyles?" Identifying the sustainable consumer. *Futures, 37*, 481–504.

Gong, Z. H., & Chu, H. (2022). Seeing risks or solutions: Psychological distance and ecological worldview moderated the effect of disgust images on attention to environmental messages. *SAGE Open, 12*, 21582440221103856.

Gupta, S., & Srivastav, P. (2016). An exploratory investigation of aspirational consumption at the bottom of the pyramid. *Journal of International Consumer Marketing, 28*(1), 2–15.

Hammond, A. L., & Prahalad, C. K. (2004). Selling to the poor. *Foreign Policy*. https://doi.org/10.2307/4147574

Hoffmann, R., Muttarak, R., Peisker, J., & Stanig, P. (2022). Climate change experiences raise environmental concerns and promote Green voting. *Nature Clinical Practice Endocrinology & Metabolism, 12*, 148–155.

Hosta, M., & Zabkar, V. (2021). Antecedents of environmentally and socially responsible sustainable consumer behavior. *Journal of Business Ethics, 171*, 273–293.

Hwang, J., Lee, J., Kim, J. J., & Sial, M. S. (2021). Application of internal environmental locus of control to the context of eco-friendly drone food delivery services. *Journal of Sustainable Tourism, 29*, 1098–1116.

Jagadale, S. R., Kadirov, D., & Chakraborty, D. (2018). Tackling the subaltern quandary: Marketing systems of dignity. *Journal of Macromarketing, 38*(1), 91–111. https://doi.org/10.1177/0276146717740680

Jaikumar, S., Singh, R., & Sarin, A. (2018). 'I show of, so I am well of': Subjective economic well-being and conspicuous consumption in an emerging economy. *Journal of Business Research, 86*, 386–393.

Jaiswal, A. K., & Gupta, S. (2015). The influence of marketing on consumption behavior at the bottom of the pyramid. *Journal of Consumer Marketing, 32*(2), 113–124. https://doi.org/10.1108/JCM-05-2014-0996

Jiao, J., & Zhao, J. (2023). Individualism, collectivism, and allocation behavior: Evidence from the ultimatum game and dictator game. *Behavioral Science (Basel), 13*(2), 169. https://doi.org/10.3390/bs13020169

Kassinis, G., Panayiotou, A., Dimou, A., & Katsifaraki, G. (2016). Gender and environmental sustainability: A longitudinal analysis. *Corporate Social Responsibility and Environmental Management, 23*, 399–412.

Kim, Y., & Choi, S. M. (2005). Antecedents of green purchase behavior: An examination of collectivism, environmental concern, and perceived consumer effectiveness. *Advances in Consumer Research, 32*, 592–599.

Kleanthous, A. (2011, July 6). Simply the best is no longer simple. Raconteur. http://np.netpublicator.com/netpublication/n104448949 (accessed January 7, 2023).

Kolk, A., Rivera-Santos, M., & Rufín, C. (2014). Reviewing a decade of research on the "base/bottom of the pyramid" (BOP) concept. *Business & Society, 53*(3), 338–377.

Korsunova, A., Halme, M., Kourula, A., Levanen, J., & Toivanen, M. L. (2022). Necessity-driven circular economy in low-income contexts: How informal sector practices retain value for circularity. *Global Environmental Change, 76*, Article 102573. https://doi.org/10.1016/j.gloenvcha.2022.102573

Kubacki, K., Dietrich, T., & Rundle-Thiele, S. (2017). Segmentation in social marketing: Why we should do it more often that we currently do. In T. Dietrich, S. Rundle-Thiele, K. Kubacki (Eds.), *Segmentation in social marketing: Process, methods and application* (pp. 1–6). Springer Singapore.

Lee, J., Taherzadeh, O., & Kanemoto, K. (2021). The scale and drivers of carbon footprints in households, cities and regions across India. *Global Environmental Change, 66*, Article 102205. https://doi.org/10.1016/j.gloenvcha.2020.102205

Lee, Y.-J., Haley, E., & Yang, K. (2019). The role of organizational perception, perceived consumer effectiveness and self-efficacy in recycling advocacy advertising effectiveness. *Environmental Communication, 13*, 239–254.

Litvine, D., & Wüstenhagen, R. (2011). Helping "light green" consumers walk the talk: Results of a behavioural intervention survey in the Swiss electricity market. *Ecological Economics, 2011*(70), 462–474.

Lockrey, S., Brennan, L., Verghese, K., Staples, W., & Binney, W. (2018). Enabling employees and breaking down barriers: Behavioural infrastructure for pro-environmental behaviour. *Research handbook on employee pro-environmental behaviour* (pp. 313–345). Edward Elgar.

Lockrey, S., Brennan, G., Verghese, K., Staples, S., & Binney, W. (2023). Behavioural infrastructures to interrupt routines and support the development of new habits in sustainable consumption. *Sustainability, 16*(7), Article 3023. https://doi.org/10.3390/su16073023

London, T., & Hart, S. L. (2004). Reinventing strategies for emerging markets: Beyond the transnational model. *Journal of International Business Studies, 35*(5), 350–370.

Mathur, M., Mehta, R., & Swami, S. (2020). Developing a marketing framework for the bottom of the pyramid consumers. *Journal of Advances in Management Research*. https://doi.org/10.1108/JAMR-01-2020-0015

McCarty, J. A., & Shrum, L. J. (1994). The recycling of solid wastes: Personal values, value orientations, and attitudes about recycling as antecedents of recycling behavior. *Journal of Business Research, 30*(1), 53–62. https://doi.org/10.1016/0148-2963(94)90068-X

McEvoy, J. III (1972). The American concern with the environment. In W. B. Burch, Jr., C. H. Check, & L. Taylor (Eds.), *Social behaviour, natural resources and the environment*, Harper and Row.

Mishra, R., Singh, R., & Jaikumar, S. (2021). Executive functions of BoP consumers: Research propositions, conceptual framework and implications for marketing strategies for BoP markets. *Journal of Global Marketing, 34*(4), 249–269.

Mohammed, A. S., Graham, E., & Dary, S. K. (2022). Rising expectations and dying hopes: Local perceptions of oil and gas extraction in Ghana. *Energy Research and Social Science, 88*, Article 102529.

Moisander, J. (2007). Motivational complexity of green consumerism. *International Journal of Consumer Studies, 31*(4), 404–416.

Moschis, G. P., Mathur, A., & Shannon, R. (2020). Toward achieving sustainable food consumption: Insights from the life course paradigm. *Sustainability, 12*, 5359.

Mukherjee, S., Datta, B., & Paul, J. (2020). The phenomenon of purchasing second-hand products by the BOP consumers. *Journal of Retailing and Consumer Services, 57*, Article 102189.

Naatu, F., Alon, I., Amoako G., & Mornah, D. (2023). Pro-environmental values' and consumer behavior in base of the pyramid market: Ghana. *Competitiveness Review: An International Business Journal, 34*(4), 807–831.

Papadopoulou, M., Papasolomou, I., & Thrassou, A. (2022). Exploring the level of sustainability awareness among consumers within the fast-fashion clothing industry: A dual business and consumer perspective. *Competitiveness Review: An International Business Journal, 32*(3), 350–375.

Phan-Le, N. T., Brennan, L., & Parker, L. (2024). An integrated model of the sustainable consumer. *Sustainability, 16*, 3023. https://doi.org/10.3390/su16073023

Prahalad, C. K. (2004). *The fortune at the bottom of the pyramid: Eradicating poverty through profits*. Wharton School Publishing.

Preston, F., & Lehne, J. (2017). *A wider circle? The circular economy in developing countries* (p. 5). Chatham House Briefing.

Rathore, P., Saha, E., Chakraborty, S., & Tiwari, A. K. (2022). Assessing impact of consumer perceived CSR on consumer attitude and purchase behaviour in retail segment: A stakeholder theory perspective. *Society and Business Review, 18*(2), 1746–5680. https://doi.org/10.1108/SBR10-2021-0207

Rausch, T. M., & Kopplin, C. S. (2021). Bridge the gap: Consumers' purchase intention and behavior regarding sustainable clothing. *Journal of Cleaner Production, 278*, Article 123882.

Roberts, J. A. (1996). Green consumers in the 1990s: Profile and implications for advertising. *Journal of Business Research, 36*(3), 217–232.

Roberts, J. A., & Bacon, D. R. (1997). Exploring the subtle relationships between environmental concern and ecologically conscious consumer behavior. *Journal of Business Research, 1997*(40), 79–89.

Sheth, J. N. (2011). Impact of emerging markets on marketing: Rethinking existing perspectives and practices. *Journal of Marketing, 75*(4), 166–182. https://doi.org/10.1509/jmkg.75.4.166

Srivastava, A., Mukherjee, S., & Jebarajakirthy, C. (2020). Aspirational consumption at the bottom of pyramid: A review of literature and future research directions. *Journal of Business Research, 110*, 246–259.

Straughan, R. D., & Roberts, J. A. (1999). Environmental segmentation alternatives: A look at green consumer behaviour in the new millennium. *Journal of Consumer Marketing, 16*(6), 558–575.

Tansan, B., Lang, N., Meyer, M., Gökbulut, A., Ivers, L., Hutchinson, R., Santamarta, S., Azevedo, D., & Chan, T. (2023, March). *The sustainability imperative in emerging markets*. Boston Consulting Group. https://www.bcg.com/publications/2023/global-esg-challengers-the-sustainability-imperative-in-emerging-markets

Triandis, H. C. (1993). Collectivism and individualism as cultural syndromes. Cross-cultural research. *The Journal of Comparative Social Science, 27*(3–4), 155–180. https://doi.org/10.1177/106939719302700301

Trudel, R. (2019). Sustainable consumer behavior. *Consumer Psychological Review, 2019*(2), 85–96.

Venugopal, S., & Viswanathan, M. (2017). The subsistence marketplaces approach to poverty: Implications for marketing theory. *Marketing Theory, 17*(3), 341–356.

Verplanken, B., & Orbell, S. (2003). Reflections on past behavior: A self-report index of habit strength. *Journal of Applied Social Psychology, 33*, 1313–1330.

Verplanken, B., Roy, D., & Whitmarsh, L. (2018). Cracks in the wall: Habit discontinuities as vehicles for behaviour change. In B. Verplanken (Ed.), *The psychology of habit: Theory, mechanisms, change, and contexts* (pp. 189–205). Springer International Publishing.

Verplanken, B., Walker, I., Davis, A., & Jurasek, M. (2008). Context change and travel mode choice: Combining the habit discontinuity and self-activation hypotheses. *Journal of Environmental Psychology, 28*, 121–127.

Weigel, R., & Weigel, J. (1978). Environmental concern: The development of a measure. *Environment and Behavior, 10*, 3–15.

Wood, W., & Rünger, D. (2016). Psychology of habit. *Annual Review of Psychology, 2016*(67), 289–314.

Zaki, Y., Chen, J., Pötsch, T., Ahmad, T., & Subramanian, L. (2014). Dissecting web latency in Ghana. In *Proceedings of the 2014 Conference on Internet Measurement Conference* (pp. 241–248).

Zaremohzzabieh, Z., Ismail, N., Ahrari, S., & Abu Samah, A. (2021). The effects of consumer attitude on green purchase intention: A meta-analytic path analysis. *Journal of Business Research, 132*, 732–743.

Zheng, X., Baskin, E., & Peng, S. (2018). Feeling inferior, showing of: The effect of nonmaterial social comparisons on conspicuous consumption. *Journal of Business Research, 90*, 196–205.

Ziesemer, F., Hüttel, A., & Balderjahn, I. (2021). Young people as drivers or inhibitors of the sustainability movement: The case of anti-consumption. *Journal of Consumer Policy, 2021*(44), 427–453.

Zimmer, M. R., Stafford, T. F., & Stafford, M. R. (1994). Green issues: Dimensions of environmental concern. *Journal of Business Research, 30*(1), 63–74.

Mahima Mathur is an Associate Professor of Marketing at Welingkar Institute of Management, Mumbai, India. Her Ph.D. and MPhil are in the area of consumer behaviour. Prior to joining academics Mahima has worked for organizations such as American Express in leadership positions. She has contributed to social projects at the Tata Institute of Social Sciences, Mumbai, India. Her research interests include bottom of the pyramid, consumer behaviour and social marketing. She has published several research papers and case studies in leading international journals. Her teaching interests include consumer behaviour, marketing management and integrated marketing communications.

The manufacturer's authorised representative in the EU is Springer Nature Customer Service Centre GmbH, Europaplatz 3, 69115 Heidelberg, Germany. If you have any concerns regarding our products, please contact ProductSafety@springernature.com

Printed and bound by CPI Group (UK) Ltd, Croydon, CR0 4YY
01/10/2026
02225915-0001